T0340080

The New Era of Precision Medicine

The New Era of Precision Medicine

What it Means for Patients and the Future of Healthcare

Edited by

MOHAMAD BYDON M.D.

Mayo Clinic, Rochester, MN, United States

ELSEVIER

ACADEMIC PRESS

An imprint of Elsevier

Academic Press is an imprint of Elsevier
125 London Wall, London EC2Y 5AS, United Kingdom
525 B Street, Suite 1650, San Diego, CA 92101, United States
50 Hampshire Street, 5th Floor, Cambridge, MA 02139, United States
The Boulevard, Langford Lane, Kidlington, Oxford OX5 1GB, United Kingdom

Copyright © 2024 Elsevier Inc. All rights reserved, including those for text and data mining, AI training, and similar technologies.

No part of this publication may be reproduced or transmitted in any form or by any means, electronic or mechanical, including photocopying, recording, or any information storage and retrieval system, without permission in writing from the publisher. Details on how to seek permission, further information about the Publisher's permissions policies and our arrangements with organizations such as the Copyright Clearance Center and the Copyright Licensing Agency, can be found at our website: www.elsevier.com/permissions.

This book and the individual contributions contained in it are protected under copyright by the Publisher (other than as may be noted herein).

Notices

Knowledge and best practice in this field are constantly changing. As new research and experience broaden our understanding, changes in research methods, professional practices, or medical treatment may become necessary.

Practitioners and researchers must always rely on their own experience and knowledge in evaluating and using any information, methods, compounds, or experiments described herein. In using such information or methods they should be mindful of their own safety and the safety of others, including parties for whom they have a professional responsibility.

To the fullest extent of the law, neither the Publisher nor the authors, contributors, or editors, assume any liability for any injury and/or damage to persons or property as a matter of products liability, negligence or otherwise, or from any use or operation of any methods, products, instructions, or ideas contained in the material herein.

ISBN: 978-0-443-13963-5

For Information on all Academic Press publications
visit our website at https://www.elsevier.com/books-and-journals

Publisher: Mara Conner
Acquisitions Editor: Sonnini Yura
Editorial Project Manager: Susan Ikeda
Production Project Manager: Jayadivya Saiprasad
Cover Designer: Greg Harris

Typeset by MPS Limited, Chennai, India

Working together
to grow libraries in
developing countries

www.elsevier.com • www.bookaid.org

Contents

11. The financial burden of precision medicine   **229**

Sufyan Ibrahim, Karim Rizwan Nathani and Mohamad Bydon

List of contributors

Edward Abrahams
Personalized Medicine Coalition, Washington, DC, United States

Ourania Argyropoulou
Department of Pathophysiology, School of Medicine, National and Kapodistrian University of Athens, Athens, Greece

Florian Brockmann
Division of Immunology, Department of Biology, University of Konstanz, Konstanz, Germany

Mohamad Bydon
Department of Neurologic Surgery, Mayo Clinic, Rochester, MN, United States; Mayo Clinic Neuro-Informatics Laboratory, Mayo Clinic, Rochester, MN, United States

Claudio Carini
School of Cancer & Pharmaceutical Sciences, Faculty of Life Sciences & Medicine, New Hunt's House, King's College London, Guy's Campus, London, United Kingdom; Biomarkers Consortium, Foundation of the National Institute of Health, Bethesda, MD, United States

Loukas G. Chatzis
Department of Pathophysiology, School of Medicine, National and Kapodistrian University of Athens, Athens, Greece

Ashis Dhar
Mayo Clinic Neuro-Informatics Laboratory, Mayo Clinic, Rochester, MN, United States; Department of Neurologic Surgery, Mayo Clinic, Rochester, MN, United States

Gregory J. Downing
Innovation Horizons, Inc., Washington, DC, United States; School of Health, Georgetown University, Washington, DC, United States

Abdul Karim Ghaith
Mayo Clinic Neuro-Informatics Laboratory, Mayo Clinic, Rochester, MN, United States; Department of Neurologic Surgery, Mayo Clinic, Rochester, MN, United States

Marc Ghanem
Gilbert and Rose-Marie Chagoury School of Medicine, Lebanese American University, Beirut, Lebanon

Sufyan Ibrahim
Department of Neurologic Surgery, Mayo Clinic, Rochester, MN, United States; Mayo Clinic Neuro-Informatics Laboratory, Mayo Clinic, Rochester, MN, United States

Konstantinos Katsos
Mayo Clinic Neuro-Informatics Laboratory, Mayo Clinic, Rochester, MN, United States; Department of Neurologic Surgery, Mayo Clinic, Rochester, MN, United States

Subhan Khan
UK Comprehensive Epilepsy Program, University of Kentucky, Lexington, KY,
United States; Department of Neurosurgery, Kentucky Neuroscience Institute (KNI),
University of Kentucky, Lexington, KY, United States; Westchester Medical Centre,
Valhalla, NY, United States

Matthias Kohl
Medical and Life Sciences Faculty, Furtwangen University, Villingen-Schwenningen,
Germany; Institute of Precision Medicine, Furtwangen University, Villingen-
Schwenningen, Germany

Longqi Liu
BGI Research, Hangzhou, Hangzhou, China

Alexios-Fotios A. Mentis
BGI Genomics, Shenzhen, China

Farhan A. Mirza
UK Comprehensive Epilepsy Program, University of Kentucky, Lexington, KY,
United States; Department of Neurosurgery, Kentucky Neuroscience Institute (KNI),
University of Kentucky, Lexington, KY, United States

F.M. Moinuddin
Mayo Clinic Neuro-Informatics Laboratory, Mayo Clinic, Rochester, MN, United States;
Department of Neurologic Surgery, Mayo Clinic, Rochester, MN, United States

Karim Rizwan Nathani
Department of Neurologic Surgery, Mayo Clinic, Rochester, MN, United States; Mayo
Clinic Neuro-Informatics Laboratory, Mayo Clinic, Rochester, MN, United States

Panagiota Palla
Department of Pathophysiology, School of Medicine, National and Kapodistrian
University of Athens, Athens, Greece

Konstantinos Panagiotopoulos
Department of Pathophysiology, School of Medicine, National and Kapodistrian
University of Athens, Athens, Greece

Lisa S. Parker
Center for Bioethics & Health Law, University of Pittsburgh, Pittsburgh, PA,
United States; Department of Human Genetics, University of Pittsburgh, Pittsburgh, PA,
United States

Filip Paskali
Medical and Life Sciences Faculty, Furtwangen University, Villingen-Schwenningen,
Germany; Institute of Precision Medicine, Furtwangen University, Villingen-
Schwenningen, Germany

Donna M. Roscoe
Division of Molecular Genetics and Pathology, Office of In Vitro Diagnostics, Office of
Product Evaluation and Quality Center, Center for Devices and Radiological Health,
U.S. Food and Drug Administration, Silver Spring, MD, United States

Weronika Schary
Medical and Life Sciences Faculty, Furtwangen University, Villingen-Schwenningen, Germany; Institute of Precision Medicine, Furtwangen University, Villingen-Schwenningen, Germany

Attila A. Seyhan
Laboratory of Translational Oncology and Experimental Cancer Therapeutics, Warren Alpert Medical School, Brown University, Providence, RI, United States; Department of Pathology and Laboratory Medicine, Warren Alpert Medical School, Brown University, Providence, RI, United States; Joint Program in Cancer Biology, Lifespan Health System and Brown University, Providence, RI, United States; Legorreta Cancer Center, Brown University, Providence, RI, United States

Jonathan Simantzik
Medical and Life Sciences Faculty, Furtwangen University, Villingen-Schwenningen, Germany; Institute of Precision Medicine, Furtwangen University, Villingen-Schwenningen, Germany

Athanasios G. Tzioufas
Department of Pathophysiology, School of Medicine, National and Kapodistrian University of Athens, Athens, Greece

Preface

Precision medicine represents a continuously shifting paradigm that defines our approach to modern healthcare provision. It encompasses a comprehensive understanding of lifestyle, behavioral, genetic, and environmental determinants of health at an individual level and the intricate interplay between these factors.

This book is an exploration of the rapidly evolving field of precision medicine and aims to engage both healthcare professionals and individuals seeking to understand this discipline. The knowledge shared within will shed light on the past, present, and future of personalized healthcare, exploring its transformative potential, ethical complexities, and global impact. From the sophisticated methodologies of big data and artificial intelligence to the compelling narratives of success and the invaluable lessons learned from challenges, this book stands as a tribute to the remarkable progress we have achieved and the paths that lie ahead.

Moreover, the book endeavors to examine the interactions between precision healthcare and both medical and surgical interventions, highlighting the transformative potential and limitations of this novel approach. We extend our heartfelt gratitude to our colleagues and fellow contributors, whose unwavering support has immensely enriched this project. Their collective efforts and dedication have shaped this book into a testament to our shared vision of a more customized healthcare practice worldwide. We hope this book contributes a positive change in our constant pursuit of modernizing healthcare through precision medicine.

Acknowledgment

We extend our sincere appreciation to our colleagues for their invaluable contributions and tireless support.

Introduction

In a time defined by rapid technological advancements and unprecedented scientific breakthroughs, medicine is undergoing a metamorphosis that goes beyond traditional approaches. Precision medicine, a revolutionary concept, is challenging conventional standards by setting the individual at the center of healthcare performance. This transition is redefining diagnosis, treatment, and prevention, promising a future where medical interventions are optimized to suit the unique biological, genetic, and environmental makeup of each patient.

We embark on a journey through the origins of precision medicine, tracing its roots in the remarkable discoveries of genetics and genomics. From the pioneering efforts of scientists such as Gregor Mendel and Rosalind Franklin to the revolutionary breakthroughs of the Human Genome Project, we witness the unfolding of the era of the individual-based practice of medicine. Furthermore, this book discusses an array of multiple compelling case studies that highlight the practical applications of precision medicine across various medical specialties. We witness how precision oncology has revolutionized cancer treatment by targeting specific genetic mutations, leading to improved outcomes and better quality of life for patients. We also explore the possibilities of precision practice in other fields, such as epilepsy management in neurologic surgery, illustrating the potential for interventions tailored to the needs of each patient.

Throughout our journey, we had the privilege of interacting with numerous experts, researchers, and practitioners working tirelessly to promote and offer precision medicine. The invaluable exchange of knowledge and experiences with these individuals solidified our belief in the power of collaboration and knowledge sharing. It also became evident that bringing together the collective expertise of esteemed contributors in a single comprehensive volume would not only showcase the multifaceted aspects of precision medicine but also serve as a catalyst for further advancements.

The book also highlights the frequent challenges associated with the practice of precision healthcare, including ethical considerations, privacy protection, and the growing need for interdisciplinary collaboration. We discuss the complex balance of availing the opportunities of benefit with the risks of potential harm to the patients. We hope that this book serves as a guide to all those who seek to comprehend the practice of precision medicine that continues to redefine the boundaries of modern healthcare.

CHAPTER 1

On the modern evolution of personalized medicine

Edward Abrahams[1] and Gregory J. Downing[2,3]
[1]Personalized Medicine Coalition, Washington, DC, United States
[2]Innovation Horizons, Inc., Washington, DC, United States
[3]School of Health, Georgetown University, Washington, DC, United States

Personalized medicine is the future of healthcare, and throughout the decades to come, we will be realizing that future at an increasingly rapid pace. In the two decades that have passed since the completion of the Human Genome Project, individuals have raised the question as to when personalized medicine will become the norm and when they will be able to benefit from it. Here in this chapter, we will highlight the current application of personalized medicine in healthcare delivery. The selection of a medical oncology regimen that has been prescribed based on a genetic test result, personal health record use to monitor the accuracy of diabetes control and insulin management, and the use of remote monitoring devices to detect and treat arrhythmias are examples of how personalized healthcare is currently used in practice. These represent just a few examples of how patients experience personalized medicine in a wide array of dimensions in an emerging healthcare system that promises future innovations to improve outcomes and quality of life.

This chapter is developed from a historical perspective to demonstrate how scientific knowledge derived from the genetic understanding of disease has overcome barriers in clinical medicine through the arc of time. While the impact has led to a radical transformation of the management of cancer, cardiovascular disease, immunology, and many disciplines of healthcare, this path has had many challenges concerning the integration of personalized medicine into the existing paradigm of healthcare. We recognize that understanding these challenges is important and that new approaches are needed to further propel the practice of personalized medicine toward a future where all patients will experience its benefits without limitations.

It was at a crucial time in the early history of personalized medicine that an aspirational "call to action" was heard for this healthcare concept

The New Era of Precision Medicine
DOI: https://doi.org/10.1016/B978-0-443-13963-5.00009-1

© 2024 Elsevier Inc.
All rights reserved.

that the scientific and medical community vigorously pursued as "the right medicine, at the right time, for the right person" [1]. These voices for change were directed to Congress and government agencies for R&D investment and regulatory policy changes; to investors to seize the human genome as a high-impact/high-risk investment opportunity with important social and economic returns; to researchers as an incentive to pursue an unbridled opportunity for discovery; and to patients to participate in research and advocate for personalized medicine therapies that could benefit family and friends.

The arrival of targeted molecular therapy at the end of the 20th century represented a massive pivot in strategy from the historical approach to medical practice represented by diagnostic and treatment regimens that were uniform for most patients with the same diagnosis. In this system, success was determined by patient response to the prescribed therapy of choice. While differences in outcomes were widely acknowledged, the "art and science" of medical management was largely defined by evidence derived from large population-based clinical trials and care plans that refined management regimens based on trial-and-error approaches. The safety of medications was also emerging as a prominent issue when rare but serious adverse events and medication errors occurred. During the same time period, there were prominent concerns about the future of biomedical research; as the pipeline of new molecular entities (NME) was dwindling, the costs associated with large randomized clinical trials and drug development were soaring, and the duration of time from discovery to market entry was nearly 15 years.

The paradigm of 21st-century healthcare that we now face is the result of new models of care and payment aimed at achieving value over volume, equitable access to novel therapies, and continuous and connected care strategies for disease management, prevention, wellness, and environmental risk management. The hallmark of value-based care is that the incentives for payment are aligned to reward quality of care over the volume of services. Each facet of this image of our future healthcare system has at its core the common characteristic of precision in describing a clinical action that is undertaken with a discrete and defined biological meaning. It is this principle that at its core differentiates the past from the future in providing more effective, preventative, and safer healthcare that we have today.

Now, the signals are clear that the genome and genomic research has had a prominent impact on filling the drug development pipeline.

Various databases have shown that there are thousands of active molecular targets for drug development that continue to grow at a high rate Zhou et al. [2,3] Over the past 5 years, therapies developed on the basis of disease biomarkers and clinically selected or monitored by test results have consistently represented over one-third of the annual new drug entities approved by the U.S. Food and Drug Administration (FDA), and the total number approved has doubled during this time period Personalized Medicine Coalition [4]. In summary, the key point to underscore is that the foundation of personalized medicine is sound, and the future expansion of its application is clearly in sight. The evidence is also clear that we are experiencing a continuum of change in the healthcare delivery system that is shifting priorities and realigning services that appear to favor the opportunities that are innovative and equitable, including personalized medicine.

As we look back through the lens of time, we see that the human genome era brought with it a wave of new scientific approaches and technologies that have led to a characterization of diseases differently than the previous ritual based on symptoms, physiologic responses, and pathologic signs. These newly defined cellular processes (such as the methylation of a nucleotide) and identifiable molecular pathways (for example, a regulated protein kinase) provide a new lexicon to classify or describe diseases and themselves to the definition of the uniqueness of the individual patient's biological presentation of disease. These new descriptive features of disease are now commonly included in healthcare quality measures, coding standards, and healthcare payment policies. Over the years, researchers armed with high-fidelity (highly accurate, specific, and reproducible) analytic tools and immense computing power unleashed the basic codes of biological understandings of disease. The result has now been transformed into discrete clinical reference resources that guide clinical treatment pathways based on the uniqueness of a patient's biology. Thus, nearly all aspects of the healthcare delivery system are facing changes in protocols and processes to embrace this dramatic new way of diagnosing and treating the patient. Therein lies the need for some formidable changes in the education and training of clinicians to ensure that genomic information is actionable in appropriate clinical settings when needed. Overall, the transition of the biomedical R&D engine and delivery system from a one-size-fits-all to a customization pathway that is emerging for individual treatments for common and rare diseases has been as challenging as anticipated Deverka et al. [5]. Yet, today, clinicians and medical experts have a staggering array of tools at hand for disease-modifying capabilities that were unimaginable a generation ago.

In this chapter, we approach the future of personalized medicine from the perspective of an advocacy organization that has led and taken part in shaping much of the history of personalized medicine in the United States. Here we highlight key historical milestones and the actions taken to explain the path to today and how the map to the future is designed and cultivated to allow personalized medicine practices to flourish and benefit patients' needs. Finally, we present a summary of key bottlenecks and barriers that represent the main challenges that lie ahead.

Origins of the concept of individualized treatment on the basis of human genetic differences

Today, there is a robust, multidimensional, multisector discovery and development engine at work featuring high-throughput genomic sequencing, targeted molecular and genetic therapies, and powerful data analytic systems at work, creating a new framework for disease prevention, detection, and treatment that is focused on an individual patient's characteristics of health and disease. It is instructive to examine the milestones of discovery that established the path of an individualized response to therapy. Here we examine the historical arc of key scientific understandings about the mechanisms of drug effects and how they can elicit different biological results in individuals on the basis of genetic differences. Each of these milestones in their own way contributed strategically to new approaches to applying genomic information to product development and clinical management.

Research on the pharmacological basis for the individuality of patient responses to medicines had its beginnings in pharmacogenomics in the 1950s with the discovery of different rates of drug metabolism based on genetic polymorphisms that led to harmful clinical effects Kalow [6]. At its foundation, the science of measuring drug response and biotransformation on the basis of genetic differences in individuals was primarily aimed toward concerns about drug safety. In the 1980s, several notable pharmaceutical-biotechnology collaborations led to new drugs using genetically bioengineered principles for new medical products such as recombinant insulin, opening the door for biotechnology as a new industry. In the same decade, scientists also began to make the case that sequencing the human genome would also serve to accelerate biomedical research and provide tools to elucidate the mechanisms and causes of human disease. Proponents of the Human Genome Project claimed it would bring about a profound transformation in drug development and the practice of medicine. Even at that time, there was a common belief that genomics would become the mainstay of discovery for the pharmaceutical industry in the 21st century. The Project, which completed a map of nearly 25,000 human genes, was conducted from 1990 to 2003. The establishment of the data from the project as an open information resource was key to revolutionizing gene discovery and charting a new path to therapeutic development.

In the 1980s, genetic research by Nobel Laureates Michael Brown and Joseph Goldstein uncovered the hereditary basis for disorders of cholesterol metabolism, which subsequently led to the development of therapies based on their discovery. They discovered that the underlying mechanism for the severe form of hereditary familial hypercholesterolemia is a complete, or partial, lack of functional LDL receptors. They also discovered that in normal individuals the uptake of dietary cholesterol inhibits the cells' own synthesis of cholesterol. Consequently, the number of LDL receptors on the cell surface is reduced and leads to increased levels of cholesterol in the blood, which subsequently may accumulate in the walls of arteries, causing atherosclerosis and eventually a heart attack or a stroke. These discoveries opened the door to molecular therapies to treat atherosclerosis and the prevention of stroke and heart attacks.

Therapeutics that were designed to be targeted to specific molecules based on a genetic variation associated with a disease process began in the 1990s with the development of trastuzumab (Herceptin) for certain forms of breast cancer. This discovery represented a major milestone in the pharmaceutical industry's perspective of personalized medicine for several reasons. First, the targeting of tumor genes for specific molecular interactions led to open many new lines of discovery and therapeutic discovery. The association of the HER2/neu proto-oncogene and its role in the pathogenesis of breast cancer tumors, and the development of the anti-HER2 monoclonal antibody, trastuzumab, that was directed against the HER2 receptor represented a breakthrough in breast cancer research as the first monoclonal antibody available for the treatment of this disease. Later, HER-2 overexpression was found to be an independent adverse prognostic factor and predictor for response to both chemotherapy and endocrine agents, relating to a new approach for gene expression as a disease monitoring strategy. The third strategic breakthrough came in the clinical trials design, whereby the trials recruited only women with tumors that overexpressed HER-2 identified by immunohistochemical (IHC) staining. Subsequently, trastuzumab was hailed for pointing the way to better cancer treatment design and treatment, including proving a path forward for combining a diagnostic test with a monoclonal antibody for targeted treatment. It signaled the beginning of a new era where the determination of therapeutic efficacy of a molecular therapy could be revealed in combination with a diagnostic test—a key attribute of personalized medicine that perhaps will one day quell the era of "trial and error medicine."

Advances in novel therapeutic systems of gene therapy: the future is now

Novel targeted therapies in medical oncology emerged as mainstays in the early 2000s with imatinib and trastuzumab as monoclonal antibodies that targeted cancer cells by interrupting specific molecular pathways seen in the affected cells. With many dramatic discoveries following this path, dozens of targeted therapies now exist for cancer and are used based on the molecular signatures of cancer pathways. Many other applications in immunology, cardiovascular medicine, neurology, and others are part of the mainstay of medical management. A decade later, immunotherapy—the harnessing of the body's immune system to attack cancer cells—is rapidly emerging as a high-impact form of cancer precision therapy with widespread implications and multiple approaches to harnessing the native immune system as an activated defense that is targeted to specific molecular and genetic signatures. Briefly described here are summaries of high-impact personalized medicine technologies that offer stunning options to previously unimagined treatments for some of the most devastating human diseases and conditions.

Gene editing (CRISPR/Cas-9)

In recent years, revolutionary genetic research showed that editing of genes was feasible through a variety of molecular approaches, and recent clinical developments have sparked enormous interest in potential wide applications in inherited disease therapy. This new approach also allows for multiple gene modifications to be manipulated simultaneously and has controlled levels of gene editing activity. These technologies allow genetic material to be added, removed, or altered at specific locations in the genome with the intention of improving the function of a protein to restore or enhance its biological actions. While several approaches to genome editing have been developed, a technology system referred to as CRISPR-Cas9 (clustered regularly interspaced short palindromic repeats and CRISPR-associated protein 9) has garnered a lot of excitement for clinical use in the scientific community because it is faster, cheaper, more accurate, and more efficient than others. Further modifications of the system have enabled gene imaging labels to be used that offer ways to evaluate the DNA and RNA transcript activity with visual systems to monitor response to light signals.

The clinical application of CRISPR/Cas-9 applications for gene therapy appears on the horizon for various types of inherited diseases such as sickle cell disease, Duchenne muscular dystrophy, and certain hemophilias Wu et al. [7,8]. Important new applications of CRISPR/Cas-9 use are enabling researchers to create models of various cancers by generating mutations that diminish or enhance gene function. Therapeutic approaches to regulate cancer tumor expression through gene editing are strengthening the role of regulating T-cell immunotherapy modes of overcoming cancer processes. Nevertheless, the technology comes with significant ethical concerns, including safety and access, where attention is needed to avoid untoward social concerns and aspects of equity. Although this work has been ongoing for many years, it is viewed that this is the dawn of a new era of personalized medicine, and for that reason, we take a brief look at these genetic therapies.

Chimeric antigen receptor-T cell therapy

Chimeric antigen receptor (CAR) T-cell therapy is a revolutionary new form of personalized cancer care Sterner and Sterner [9]. CARs are engineered synthetic receptors that function to redirect lymphocytes (T-cells) to recognize and eliminate cells expressing a specific target antigen. CAR binding to target antigens expressed on cancer cell surfaces creates brisk T-cell activation and an antitumor response. CAR-T cell therapy has had wide-scale adoption for the treatment of immune disorders, oncology, and expanding into other systemic disorders, and this has ignited many new investigational uses. Implied with the CAR T-cell approach is the recognition that each patient requires a unique, customized intervention, leading to many challenges in the delivery of patient care, costs of care, and the need for long-term follow-up. Since 2017, the Food and Drug Administration has approved six CAR T-cell therapies for the treatment of blood cancers.

Therapeutic small interfering RNA (siRNA)

Molecular silencing of disease-causing genes by nucleic acid applications to modulate gene expression has been a major research endeavor since the establishment of the RNA inhibition concept in 1998. The therapeutic reality of this difficult scientific endeavor was recently achieved in the form of small interfering RNA (siRNA) by knocking down the expression of target genes in a sequence-specific way. The opening of the door to this approach for personalized medicine was taken by the FDA recently through the approved use of patisiran for hereditary amyloidogenic transthyretin amyloidosis with polyneuropathy and givosiran for the treatment of acute hepatic porphyria Hu et al. [10]. A third novel siRNA therapeutic, lumarisan, was approved by the FDA in 2020 for the treatment of primary hyperoxaluria type 1 (PH1), an inherited rare disease of glyoxylate metabolism that arises from mutations in the enzyme alanine-glyoxylate aminotransferase. The resulting deficiency in this enzyme leads to abnormally high oxalate production, resulting in calcium oxalate crystal formation and deposition in the kidney and many other tissues, with systemic oxalosis and end stage renal disease (ESRD) being common outcomes Liebow et al. [11]. More recently, the FDA has approved inclisiran injection as a treatment to be used along with diet and maximally tolerated statin therapy for adults with heterozygous familial hypercholesterolemia (HeFH) or clinical atherosclerotic cardiovascular disease (ASCVD) who require additional lowering of low-density lipoprotein cholesterol (LDL-C). In this clinical application of twice-a-year injections, siRNA has the potential for overcoming barriers to effective management issues associated with oral cholesterol management therapies Kosmos et al. [12].

Collectively, the advances in siRNA therapies among the armamentarium of gene therapy have strategic advantages over molecular antibody drugs and other small molecular therapeutics as they interfere with the base pairing with messenger RNA, whereas the others need to recognize complex spatial relational conformations of relevant proteins. Several new forms of drug delivery systems are in development to enhance the opportunity for siRNA and other RNA gene silencing modalities for other genetic diseases.

The expanding number of diagnostics using genomic technologies

As a result of the advances in genomic technologies, there has been a dramatic increase in the production of genetic tests and testing services.

The national Genetic Test Registry (https://www.ncbi.nlm.nih.gov/gtr/) was developed to guide clinicians and researchers to the sources of genetic tests. This resource currently has a listing of over 76,000 genetic tests, covering more than 76,000 tests for over 22,500 tests. GeneReviews, an international point-of-care resource for busy clinicians, provides clinically relevant and medically actionable information for inherited conditions in a standardized journal-style format, covering diagnosis, management, and genetic counseling for patients and their families. This resource is written by one or more experts on the specific condition or disease and goes through a rigorous editing and peer review process prior to publication.

Consumer genomic testing

Personal genomics had its origins in the late 2000s with the convergence of the internet and human genome sequence information being deposited in the public domain. This convergence was also fueled by the rapid sequencing of millions of single nucleotide polymorphisms (SNPs), ultimately leading to whole exome sequencing in a matter of several years Boguski [13]. An immense amount of personal interest led to the spawning of a new industry in this new healthcare sector often referred to as the direct-to-consumer (DTC) genomic marketplace, which emerged as a way of offering consumers access to genetic information without requiring the involvement of a healthcare provider. The introduction of microarray genotyping platforms with hundreds of thousands of SNPs is likely to facilitate the development of more powerful algorithms to explicitly test for more distant genealogical relationships between individuals. The most widely recognized example of this approach to consumer genomics is 23andMe, the unicorn technology company in personal genomics. Founded by Silicon Valley technology icon Anne Wojcicki in 2006, the company emerged from a consumer model focused on a company aligned with consumer interests in heredity, health, and disease to a mega-database organization with detailed analysis to support risk assessment Hayden [14]. After a challenge to their marketing approach to disease risk assessment was issued in 2013, the company received FDA approval in 2015 for the marketing of genetic risk assessment for 10 diseases based on polymorphisms. As consumer interest in genetic risk assessment has grown, the expansive database enabled the company to begin supporting drug development driven by a consumer database supporting genetic and phenotypic discovery at a massive scale Brodwin [15]. The story of 23andMe opens a new path for the applications of genomics driven by an unprecedented resource enabled by the unbridled interest of the public in their health. Genomics has provided a new platform for which consumer interests in health are met. Among the more attractive features of these DTC platforms is the feature of periodic updates delivered to patients as the genome association patterns of mutations grow at a rapid pace.

While public interest in DTC services has continued to grow, their offerings have not been without controversy from the perspectives of ethics and clinical utility Helgeson and Stefansson [16]. One frequent concern is that the risk conferred by each variant used in DTCs tends to be low (odds ratio <2) and the accuracy in predicting the risk of disease for individuals (i.e., their clinical validity) will also be low. Secondly, some are concerned about the limited capacity of consumers to understand and cope with disease risk estimates from tests. Underlying these concerns are patronizing views that information about disease risk is dangerous to the general public unless mediated in person by medical experts.

Global public R&D investments in personalized medicine continues

In the years following the completion of the Human Genome Project, there has continued to be strong international public support in funding science across the spectrum of basic, translational, clinical, and outcomes research in personalized medicine. The United States Precision Medicine Initiative and the European Union's (EU's) "Personalized Medicine" (PerMed) were announced in 2015, and the International Consortium for Personalised Medicine (ICPerMed) was announced in 2016 Lee and Kim [17]. The US Precision Medicine Initiative is a long-term research endeavor to develop a new model of individualized care with a USD 9269 million investment from 2016 to 2020 [18,19].

In the past three years, major efforts across governments have been undertaken during the Covid pandemic to leverage investments in innovation. Rapid shifts in government policies associated with the pandemic have yielded new paradigms for personalized medicine in clinical practice. Uses of remote monitoring systems, telehealth for patient–provider communications, and continuous physiologic monitoring have gone hand in hand with molecular therapies in monitoring homeostasis, thereby saving scarce resources for inpatient care for more complex patients DeMerle et al. [20,21].

There is strong evidence to underpin the importance of government funding to support collaborations among R&D systems, large-scale systems, and supply chain networks that will serve to improve the adoption of PM by solving operational problems. This includes standards development and implementation, certification of laboratory assays and analytic tools, electronic health record systems, and decision support tools Yiu et al. [22,23]. Over the last decade, many countries have established research goals for personalized medicine and identified key priorities for overcoming barriers to PM adoption and value. As a result, areas such as inequality in medical services, health outcomes, and public health are cast in light for further policy examination. Frequently, multinational collaborations for research funding allocations are prioritized in disease-specific areas where high public health costs are realized such as in cancer, heart disease, diabetes, and degenerative brain diseases.

Key roles of the U.S. Food and Drug Administration in advancing personalized medicine

The integration of genomic testing and therapeutics into medical practice has nearly a quarter century of engagement in regulatory dimensions of practice focused on the FDA. In parallel with the scientific advancements in technology and the basic scientific understanding of the roles that genomics, proteomics, metabolomics, epigenetics, and other related sciences play, a continuum of policy and programmatic strategies have been required to embrace this emerging science. Although the arrival of molecularly targeted therapies did not achieve mainstream adoption as a drug development scheme until the new millennium, academic researchers and pharmaceutical and biotechnology industries were emerging with a focus on a foundational approach to disease management.

Among the early steps that the FDA addressed in regulatory sciences for personalized medicine was a focus on pharmacokinetic and pharmacodynamic (PK/PD) modeling that provided new avenues of assessing the efficacy and safety testing of novel medical entities Rowland et al. [24]. The translation of pharmacogenomic sciences from laboratory animal testing to human safety assessment was an early step toward using genomic-based technologies in treatment selection, dose modification, and clinical efficacy management. In the early 2000s, the FDA pursued several key initiatives to increase the use of biomarkers and surrogate end points in clinical studies to speed up decision-making and the use of these technologies in guiding regulatory reviews Lesko and Atkinson [25]. In 2003, FDA leadership recognized that a coordinated effort for data review was needed to advance the clinical application of pharmacogenomic data and issued a draft guidance seeking standardized data from sponsors Savage [26]. At this time, there were widespread concerns about the drug development pipeline with 2004 marking a 20-year low in NMEs approved by the FDA. The costs of drug development, a series of market withdrawals, and the time for regulatory approval all loomed as major challenges to the pathway to new therapies. In 2004, the FDA initiated its Critical Path Initiative with the intent of modernizing drug development by incorporating recent scientific advances, such as genomics and advanced imaging technologies, into the process. An important part of the initiative was the use of public-private partnerships and consortia to accomplish the needed research to underpin personalized medicine (Woodcock & Woosley, The FDA Critical Path Initiative and its influence on new drug development, Woodcock and Woosley) [27]. Among the steps taken was the creation of the Critical Path Institute. The importance of a systematic approach to addressing the regulatory science of personalized medicine was undertaken as a joint effort between the National Institutes of Health (NIH) and FDA. The Agencies' leadership recognized that realizing personalized medicine would require addressing scientific challenges, such as determining which genetic markers have the most clinical significance, limiting the off-target effects of gene-based therapies, and conducting clinical studies to identify genetic variants that are correlated with a drug response Hamburg and Collins [28]. A review of the FDA strategies that initiated the regulatory science shows the large effort that was undertaken to address medical reviewer needs, database development, guidance on submission practices for pharmacogenomics, labeling practices for therapeutic monitoring and drug—drug interactions affecting biotransformation, and many more practice changes Lesko and Zineh [29]. They recognized policy challenges, such as finding a level of regulation for genetic tests that could protect patients while encouraging innovation. With the rise of federal investments associated with this initiative and the American Recovery and Reinvestment Act of 2009, a wide array of projects were undertaken to speed the development of pharmacogenomics, informatics tools, clinical trials, and many components of critical infrastructure into personalized medicine translational and clinical research.

The Congressionally approved funding for the FDA has addressed personalized medicine at crucial times with legislation that supports the pharmaceutical and biotechnology industry's regulatory processes through the Prescription Drug User Fee Amendments (PDUFA). This fiscal appropriations process in PDUFA IV (2007) supported pharmacogenomic guidance development, and the FDA reauthorization in PDUFA V (2012) further enhanced personalized medicine by allowing for increased staff at the Center for Drug Evaluation and Research (CDER) to conduct medical reviews and guidance development in the area of personalized medicine Fox [30]. Another important regulatory step achieved in 2012 was the Congressionally approved "breakthrough therapy" designation program that provided incentives to the developers of NMEs that addressed testing and approval of medications that were intended to treat serious or life-threatening conditions and that preliminary evidence suggested may provide a substantial improvement over existing treatments with regard to one or more clinically significant end points. Evidence has shown that this approach to streamlining regulatory processes led to a brisk increase in submissions, especially in oncology personalized medicine therapies Darrow et al. [31].

More recently, the FDA has taken steps to address regulatory clinical study requirements for individualized genetic therapies, the so-called "n of 1" therapies that are customized for a specific patient with rare genetic disorders Kim et al. [32,33]. Recently, the FDA issued guidance for sponsors to address investigational new drug application steps in considering market approval for antisense therapies designed for n of 1 use in patients with rare diseases United States Food and Drug Administration [34].

In April 2018, the FDA issued two final guidance that recommend approaches to streamline the submission and review of data supporting the clinical and analytical validity of next-generation sequencing (NGS)-based tests. This is important given the magnitude of the clinical applications that are emerging in multiple domains of medicine. These recommendations are intended to provide an efficient and flexible regulatory oversight approach to accommodate changes as technology advances; new standards can evolve and be used to set appropriate metrics for fast-growing fields such as NGS. Similarly, as clinical evidence improves, new assertions could be supported. This adaptive approach would ultimately foster innovation among test developers and improve patients' access to these new technologies.

FDA regulatory science will likely undertake a larger role in the future of consumer genomics in the future. The context of DTC genomics is expanding rapidly with consequences in the marketplace that raise concerns regarding the intended use of the large genomic databases that become powerful knowledge sources in the commercial world. The regulatory and medical implications for genomic testing that has no evidence-based backing for clinical decision-making continue to garner concern for social, ethical, and legal implications (Au) [35]. We can anticipate that as the association of genomic data with biological characteristics beyond clinical medical applications expands, there will be ongoing societal issues associated with its use. Time will tell whether added laws and regulations will be necessary to guide the utility of personalized medicine toward an effective future.

Regulation of laboratory-developed genomic tests

The past two decades of personalized medicine have also yielded a series of policy challenges to the regulation of laboratory-developed tests (LDTs). Regulation of the medical uses of genetic tests has been an evolving and controversial element of the modern genome era. Most experts agree that the goal is to achieve a regulatory framework that balances the protection of patients from harm and promotes innovation toward disease diagnosis and therapeutic monitoring. In 2010, the U.S. General Accounting Office found that consumer testing services provided widely variable results to consumer-directed genetic tests. An advisory committee to the US Department of Health and Human Services had recommended in 2008 that the FDA exert more regulatory authority of laboratory genetic tests. Subsequently, NIH developed a genetic test registry (https://www.ncbi.nlm.nih.gov/gtr/) to include clinical data to provide a central location for the voluntary submission of genetic test information by providers. The overarching goal of the GTR is to advance public health and research into the genetic basis of health and disease. The scope includes the test's purpose, methodology, validity, evidence of the test's usefulness, and laboratory contacts and credentials.

Other regulatory advances to accelerate personalized approaches to therapy

The vast majority of genomic tests are referred to as "laboratory-developed tests" that are not FDA approved. Typically, these "LDTs" are offered as a service, and as such, they are subject to enforcement discretion by the FDA. Tests that are marketed as kits with medical claims are subject to premarket approval. LDTs performed under clinical laboratory testing regulations require Clinical Laboratory Improvement Amendments (CLIA) that are administered under the authority of the Center for Medicare and Medicaid Services (CMS). Most genomic assay kits that are marketed are reviewed by the Center for Devices and Radiological Health (CDRH) at the FDA. Over the last decade, CDRH has published guidance regarding genetic testing, including a 2018 guidance featuring collaboration between CDER and DCRH on laboratory tests that support the association between genetic information and specific medications.

Other regulatory advances to accelerate personalized approaches to therapy

Other aspects in research that have led to the acceleration of personalized medicine include enhancements to the conduct of translational research and clinical trials with has had a net effect of increasing the number of diagnositc, therapeutic and preventative innovations. In 2014, the concept of Phase 0 trials, or microdosing, was implemented. This allows for not only the selection of drug candidates to be more likely developed successfully, but also the determination of the first dose for the subsequent Phase I clinical trial. The concept of microdosing involves the use of extremely low, nonpharmacologically active doses of a drug to define the pharmacokinetic profile of the medication in human subjects. Microdosing, thus, appears as a new viable concept in the "toolbox" of drug development activity. It appears that the microdosing strategy could complement standard animal-to-human scaling, redefining the existing concept of phase I clinical research. In the future, when research methods and technology involved in Phase 0 studies become more sophisticated, human microdosing may be applied to a number of drugs developed subsequently. Another approach to clinical trial design that has benefited personalized medicine is the use of adaptive trial design.

Detailed in 2014, the FDA guidance "Adaptive Design Clinical Trials for Drugs and Biologics" encourages drug developers to expand their use of adaptive designs, and ongoing collaboration among the FDA, academia, and industry [36,37]. Adaptive design was first implemented in the I-SPY 2 breast cancer screening trial to streamline the identification of active drugs and predictive biomarkers that could be used to guide therapy Esserman and Woodcock [38]. This groundbreaking trial modeled a new adaptive approach to advance the clinical development process. Adaptive designs can overcome limitations created by the fixed structure of traditional designs. At the end of a fixed trial, it is common for researchers to regret decisions based on assumptions regarding the doses, population sample sizes, or patient allocations that were used in the study. Instead of being forced to make these pivotal decisions with limited information before the trial, an adaptive design uses accruing information to provide more relevant data to guide critical decisions throughout the development process. Data is analyzed at designated interim points, and results are used to shape future design parameters—such as doses being used, and disease indications or patient populations being studied. This flexible approach results in a more efficient use of resources as well as more informative studies.

Collectively, the regulatory paths to date have required a daunting array of legislative, rulemaking, guidance development, public–private partnerships, and substantial human and capital resources to accommodate the opportunities that personalized medicine provides. Innovation in regulatory paradigms in the personalized era will involve many challenging new approaches to genomic modifications and regulations of biological processes.

Healthcare payment system reforms impact on the clinical advancement of personalized medicine

Among the most important drivers for the adoption of personalized medicine practices is the healthcare payment system that plays an important part of the value proposition overall. Coincident with the rise of technology and services for diagnosis and treatment that constitutes the mainstay of personalized medicine practices has been a transformative change in the payment models and delivery system led mainly by the federal government, specifically the Centers for Medicare/Medicaid Services (CMS). At its core, this reform is a transition from fee-for-service (FFS) toward value-based reimbursement to hospital systems and providers. While commercial insurers and government payors such as the Veteran's Health Administration, military health programs, and state Medicaid programs are all involved in some form of payment reform and innovation, CMS has played the lead role in payment policy reforms. Recently, with the rise in the number of genetic test and personalized medicine therapies, health payment programs such as CMS and global government healthcare payors have taken a high interest. Furthermore, a personalized approach may yield substantial health benefits and slow healthcare expenditure growth Stark et al. [39,40]. Given this, governments and public institutions have recently expanded their involvement in the research and development of personalized medicine Ginsburg, Phillips [41]. In the United States, CMS plays many important roles in the implementation of personalized medicine from coding of services, national coverage systems for payment, oversight of the CLIA program described earlier, payment of graduate education training programs, integration of quality reporting programs into payment models, and innovation in payment systems. It is beyond the scope of this chapter to describe these important facets of healthcare delivery system reform, yet it is noteworthy to mention each of these dimensions. However, it is important to address the key aspects of payment reform as they relate to personalized medicine adoption.

In 2010, Congress enacted the Patient Protection and Affordable Care Act together with the Health Care and Education Reconciliation Act of 2010, known collectively as the "Affordable Care Act" (ACA). This new law transformed the U.S. healthcare system through an expansion of coverage through tax subsidies, regulated health insurance marketplaces, and expanded publicly financed government health programs, i.e., the nation's Medicaid program. Among the main drivers of reform were the widely held concerns about the trends in medical spending that were among the highest in the world, while the value of that care of this system lagged many other nations' health programs. Until recently, the vast majority of initiatives designed to lower the growth rate in health spending were aimed at replacing or adding incentives on top of the FFS system, even while FFS is often cited as the root cause of systemic misaligned incentives that reward volume over value Miller [42]. The main criticism of the FFS system stems from its failure to provide meaningful incentives for efficiency, quality, or outcomes.

In the years following the ACA enactment, CMS, through its Innovation Center, leveraged authority through the ACA to frame new payment methods to evaluate and implement alternatives to FFS, so-called alternative payment models (APMs). A wide range of APMs integrated the costs of care through risk-sharing models with providers and hospital systems to raise quality and constrain or lower costs. These APMs, such as bundled care programs and shared savings programs, provided initially for incentive payments and later financial penalties in accordance with the performance of the system. Among the challenges confronting advocates of personalized medicine practices are significant variations in payor coverage and payment options for services such as next-generation sequencing testing and gene therapy modalities. At present, federal program coverage of new tests and therapies is commonly considered in one-off processes leaving considerable uncertainty among patients, providers, and the laboratory and biotechnology industries. Commercial coverage and payment policies, while commonly following the hand of CMS, are similarly unpredictable and inconsistent across payors.

Recognizing the complexity of navigating the payment policy channels that lie ahead for those advocating for the adoption of innovative personalized medicine strategies in the care delivery system, advocates have focused on systematic policy approaches that underscore value and quality as key hallmarks of these technologies Montgomery et al. [43]. Several recent studies have been conducted to analyze current payment policies and develop recommendations for steps to be taken to alleviate bottlenecks in personalized medicine adoption. The Oncology Care Model was one of the first APMs employed and showed mixed results in improving quality and failed to show cost savings. Among the lessons learned from this early APM effort was the pressing need for more granular data and a redesign of incentives in a healthcare system that is far removed from integrated care Mullangi et al. [44]. Other approaches to address future APM models have been suggested by health policy and economists to find more equitable solutions to cost and quality that do not impede the adoption of innovative care solutions and prevent access to personalized medical care by patients Koleva-Kolarova et al. [45].

Public–private partnerships and their role in personalized medicine

In the late 1990s, in association with the human genome project, recommendations came from industry, academia, and government that addressed that the opportunity and needs to accelerate the benefits of genomics required collaboration and partnerships Wholley [46]. The NIH Biomarkers Consortium was formed in 2004 as an NIH, FDA, and private sector research collaboration led by the NIH Foundation as a nonprofit supporting the development and qualification of biomarkers for drug development and clinical care Altar [47]. The SNP Consortium was instrumental in genetic polymorphism database development beginning in 1999 that led to the SNP map of the human genome. The Critical Path Institute was established to support development of safe and effective medicines using genetic and other biomarkers in their testing. The Genetic Association Information Network (GAIN) was established by NIH, and several pharmaceutical and biotechnology companies to build a database of 18,000 genotypes across six major diseases GAIN Collaborative Research Group [48]. Soon after the completion of the human genome, multiple new collaborations to fund disease specific research were started—many funded by industry, disease advocacy organizations, philanthropists, and others. These forms of partnerships were important in many ways in spurring scientific research and policy that required new approaches to collaboration. They brought resources at a scale that was beyond any individual researcher or team working within their own domain. These resources included access to proprietary databases, highly specialized scientific expertise, and diversified the risk and opportunities across a wide base of organizations. Often these collaborations supported consensus-building activities that led to standardized methods and measures, training and education resources, and served to accelerate clinical drug development Woodcock [49].

The role advocacy in development and adoption of personalized medicine

The formation of a community of action for a particular need has been key to many transformational changes in healthcare to occur. Advocacy organizations, defined as groups that speak to or campaign for a particular cause, have been an important force for change across the biomedical research and healthcare spectrum. The broadest form of advocacy is sharing and communicating the mission of the organization to others, while direct or grassroots lobbying is directly intended to influence legislation or legal action. Examples of powerful forces of change in HIV/AIDS activism in the 1980s set in motion how patients and stakeholder organizations can drive public interest, advocate for services, stoke change among lawmakers, and inspire commercial interests to be responsive to consumer needs and interests. For example, the Juvenile Diabetes Research Foundation led a successful public policy endeavor to draw the attention of Congress for a new funding priority focused on stem cell biology research Greenstein [50]. Today, many forms of advocacy exist that are associated with specific diseases or promote science or healthcare reform more generally.

The need for advocacy in advancing progress across healthcare is complex. Unlike many aspects of the social and economic constructs of a democracy, healthcare is a very fragmented enterprise, but, collectively, it presents as a powerful social engine with enormous economic influence.

The complex association of access to care, public health, medical technologies, insurance, social care concerns, and many others form a complex web that has often found its way for politicians and government leaders to orient to or align with the themes of important interest in the public square of discourse on the role of government in this grand endeavor of personalized medicine.

The Human Genome Project was a powerful vision of scientific achievement that attracted a vast array of interest and support from academia, industry, and patient groups and had strong bipartisan policy affinity. Upon the completion in 2003, advocacy continued to address the application of the genome in healthcare through specific initiatives and policies. Important examples of advocacy include legislative achievements that addressed important barriers to the social benefits of genetic information, as was the case with the Genetic Information Non-discrimination Act of 2008 (GINA) that served as a legislative solution by protecting individuals against discrimination based on their genetic information in health coverage and in employment. Advocacy in personalized medicine was also advanced for Congressional support for the 21st Century Cures Act of 2016 that enabled substantial NIH funding for medical product development for rare diseases, modified FDA processes, updated human subject protections, and improved informed consent procedures. The NIH funding through this authority has supported large investments in personalized medicine such as the All of Us research program and the Cancer Moonshot Initiative. Many individual organizations and coalitions now include genomic and personalized medicine advocacy activities among their educational and lobbying efforts.

The Personalized Medicine Coalition (PMC) is a 501(c) nonprofit organization based in Washington, D.C., and serves as a leading advocacy organization for the cause of advancing science and medicine to advance personalized medicine. PMC, an educational and advocacy organization representing a broad spectrum of academic, industrial, patient organizations, provider, health systems, investors, and payer communities that share an interest in advancing the understanding and adoption of personalized medicine concepts and products, was launched in November 2004. The organization set forth four key goals:

- Provide leadership on public policy issues that affect personalized medicine,
- Help educate the public, policymakers, government officials, and private sector healthcare leaders about the public and personal health benefits of personalized medicine,

- Serve as a forum for identifying and informing others of those public policies that may impede the ability to deliver the promise of personalized medicine, and
- Create a structure for achieving consensus positions on crucial public policy issues and supporting changes needed to further the public interest in personalized medicine.

PMC achieves its advocacy for the community by leveraging the talents and efforts of its members and staff through its committee structures for public policy and science policy as well as ad hoc working groups. The Coalition conducts public interest surveys, workshops, and conferences geared to facilitating dialog and understanding among various audiences about the key issues affecting the implementation of personalized medicine. Also noteworthy of the Coalition's efforts are consensus policy documents that frame issues and options for addressing barriers to personalized medicine. Importantly, Congress is now among the target audiences for the Coalition's message. In February of 2020, the Personalized Medicine Coalition's educational and advocacy efforts inspired a bipartisan, bicameral group of lawmakers to launch the Congressional Personalized Medicine Caucus. The members of the caucus are committed to ensuring that healthcare policies in the United States encourage decision-makers in the public and private sectors to collaborate with patients in pursuit of prevention and treatment plans that are tailored more closely to patients' biological characteristics, circumstances, and values.

The Coalition's strategy for actionable advocacy is presented in the form of a Strategic Plan and augmented by engagement sessions and briefings with key agencies and stakeholder organizations for consensus development. Among the issues that the Coalition is addressing are payment policies for genomics, clinical workforce training in the application of genomic tests, health equity in access to personalized medicine practices, alignment of clinical trials data requirements by the FDA and CMS, data sharing practices, and the use of real-world evidence in clinical studies among others.

With nearly 20 years of public service devoted toward advancing personalized medicine, the Coalition members collectively represent an influential voice in government and public policy about the future of this modern approach to healthcare. The organization's deep knowledge of the scientific advances, policy analysis insights, and its balanced approach to the interests of patients, researchers, and other advocates is an important force for progress. As the impact of personalized medicine on all dimensions of healthcare expands, the work of advocacy organizations behind genomics and healthcare innovation will remain critical to ensure that the equity, value, and benefits the promise of personalized medicine has envisioned are fully realized.

Conclusion

Important societal issues related to equity, access, and ethics lie ahead, and addressing them through the public forum with transparency is paramount for the full potential of personalized medicine to be embraced. With a cresting wave of new technologies and care models emerging on the horizon, time is of the essence for employing new systems for evaluating the long-term impact and value of personalized medicine diagnostics and therapies. In the near term, the topic of payment and coverage of personalized medicine services will remain a top priority for advocates and policymakers to address to avoid greater gaps in care and outcomes for patients who can benefit from these interventions. As the transformation of the payment and quality system persists as a national priority, it will be crucial to adopt policies that create pathways for further innovation and clinical practice adoption of personalized medicine.

As many have stated, there is tremendous potential for personalized medicine to improve the lives of patients. The evidence for the benefits of the evolution and adoption of personalized or precision medicine can be measured by the growing number of FDA-approved therapies and the expansion to new indications of previously approved therapies. Indications are that a broad array of applications of genome sequencing tests and gene therapy options will become mainstream tools in common practice. A substantial redesign of the healthcare delivery system is necessary to fully embrace the potential in prevention, diagnosis, and treatment with personalized medicine technologies. It is important to acknowledge that while challenges remain in the areas of discovery, regulatory policy, coverage, reimbursement, and clinical adoption, personalized medicine is expected to deliver on the promise to "improve outcomes for patients and have a tremendous impact on the efficacy and efficiency of health care."

References

[1] Abrahams A. Right drug - right patient- right time: personalized medicine coalition. Clin Transl Sci 2008;1(1):11−12.
[2] EY. Beyond borders: EY biotechnology report 2022. New York City: EY; 2022.

[3] Zhou Y, Zhang Y, Lian X, Li F, Wang C, Zhu F, et al. Therapeutic target database update 2022: facilitating drug discovery with enriched comparative data of target agents. Nucleic Acids Res 2022;50(7):D1398−407.

[4] Personalized Medicine Coalition. Personalized Medicine at FDA: The Scope and Significance of Progress in 2021. Washington, DC: Personalized Medicine Coalition; 2022.

[5] Deverka PA, Doksum T, Carlson RJ. Integrating molecular medicine into the US health care system: opportunities, barriers, and policy challenges. Clin Pharmacol Ther 2007;82(4):427−34.

[6] Kalow W. Human pharmacogenomics: the development of a science. Hum Genomics 2004;1(5):375−80.

[7] Wu SS, Li QC, Yin CQ. In vivo delivery of CRISPR-Cas9 using lipid nanoparticles enables antithrombin gene editing for sustainable hemophiia A and B. Sci Adv, 2022;eabj69013.

[8] Zhang Y, Li H, Min YL. Enhanced CRISPR-Cas9 correction of muscular dystrophy in mice by a self complementary AAV delivery system. Sci Adv 2020;6(8): eaay6812.

[9] Sterner RC, Sterner RM. CAR-T cell therapy: current limitatins and potential strategies. Blood Cancer J 2021;11(69). Available from: https://doi.org/10.1038/s41408-021-00459-7.

[10] Hu B, Zhong Y, Peng L, Huang Y, Zhao Y, Liang X-J. Therapeutic siRNA: state of the art. Signal Transduct Target Ther 2020;5:101.

[11] Liebow A, Li X, Racie T, Hettinger J, Bettencourt BR, Najafian N, et al. An investigationa RNAi therapeutic targeting glycolate oxidase reduces oxalate productini in models of primary hyperoxaluria. JASN 2017;28(2):494−503.

[12] Kosmos CE, Munoz Estrella A, Sourlas A, Silverio D, Hilario E, Montan PD, et al. Inclisiran: A promising new agent for the management of hypercholesterolemia. Diseases 2018;6(3). Available from: https://doi.org/10.3390/diseases6030063.

[13] Boguski MS. Online health information retrieval by consumers and the challenge of personal genomics. In: Willard H, Ginsburg G, editors. Genomic and personalized medicine. Elsevier; 2009. p. 252−7.

[14] Hayden EC. The rise and fall and rise of 23andMe. Nature 2017;550:174−7.

[15] Brodwin E. DNA testing company 23andMe has signed a $300 million deal with a drug giant. 2018, July 25. Retrieved from https://www.businessinsider.com/dna-tsting-delete-your-data-23andme-ancestry-2018-7: Https.

[16] Helgeson A, Stefansson K. Past, present and future of directo-to-consumer genetic tests. Transl Res 2010;61−8. Available from: https://doi.org/10.31887/DCNS.2010. 12.1/ahelgason.

[17] Lee D, Kim K. Public R&D projects-based investment and collaboration framework for an overarching South Korean National STrategy of Personalized Medicine. Int J Env Res Public Health 2022;19:1291−315.

[18] National Institutes of Health. Estimates of funding for various research, condition, and disease categories (RCDC). 2021. Retrieved from National Institutes of Health: http://www.nih.gov.

[19] Ten Have H, Gordihn B. Precision in health care. Med Health Care Philos 2018;21:441−2.

[20] Kadakia K, Patel B, Shah A. Advancing digital health: FDA innovation during COVID-19. NPJ Digit Med 2020;3:161−4.

[21] DeMerle K, Angus DC, Seymour CW. Precision medicine for COVID-19. JAMA 2021;324:2041.

[22] Vicente AM, Ballensiefen W, Jonsson JI. How personalised medicine will transform healthcare by 2030: The ICPerMed vision. J Transl Med 2020;18:1−4.

[23] Yiu C, Macon-Cooney B, Fingerhut H. A research and policy agenda for the post-pandemic world. Futur Health J 2021;8:e198−203.

[24] Rowland M, Peck C, Tucker G. Physiologically-based pharmacokinetics in drug development and regulatory science. Ann Rev Pharmacology Toxicol 2011;51:45−73.

[25] Lesko L, Atkinson A. Useof biomarkers and surrogate endpoints in drug development and regulatory decisionmaking: criteria, validation, strategies. Annu Rev Pharm Toxicol 2001;41:347−66.

[26] Savage DR. FDA guidance on pharmacogenomics data submission. Nat Rev Drug Discovery 2003;2(12):937−8.

[27] Woodcock J, Woosley R. The FDA critical path initiative and its influence on new drug development. Annu Rev Med 2008;59:1−12.

[28] Hamburg MA, Collins FS. The path to personalized medicine. N Engl J Med 2010;363(22):301−4.

[29] Lesko LJ, Zineh I. DNA, drugs and chariots: on a decade of pharmacogenomics at the US FDA. Pharmacogenomics 2010;11(4):10−16.

[30] Fox JL. Rare-disease drugs boosted by new Prescription Drug User Fee Act. Nat Biotechnol 2012;30(8):733−4.

[31] Darrow JJ, Avorn J, Keselheim AS. The FDA breakhrough-drug designation - four years of experience. N Engl J Med 2018;378(15):1445−54.

[32] Kim J, Hu C, Moufawad El Achkar C. Patient-customized oligonucleotide therapy for a rare genetic disease. N Engl J Med 2019;381:1644−52.

[33] Woodcock J, Marks P. Drug regulation in the era of individualized therapies. N Eng J Med 2019;381:1678−80.

[34] United States Food and Drug Administration. IND submissions for individualized antisense oligonucleotide products: administrative and procedural recommendations guidance for sponsor-investigators. 2021, January. Retrieved November 2022, from Drugs/Guidance, Compliance, & Regulatory Information: https://www.fda.gov/drugs/guidance-compliance-regulatory-information/guidances-drugs.

[35] Sanghavi K, Cohn B, Prince AJ. Voluntary workplace genomic testing: wellness benefit or Pandora's box? npjGenom Med 2022;7(5). Available from: https://journals.sagepub.com/doi/10.1177/09636625211051964.

[36] Burt T, Yoshida K, Lappin G, Vuong L, John C, de Wildt SN, et al. Microdosing and other phase 0 clinical trials: facilitating translation in drug development. Clin Transl Sci 2016;9(2):74−88.

[37] US Food and Drug Administration. Draft guidance for industry: adaptive design clinical trials for drugs and bilogics. 2010. Retrieved from U.S. Food and Drug Administration: http://www.fda.gov/downloads/drugs/guidancecomplianceregulatoryinformation/guidancesucm2017.pdf.

[38] Esserman L, Woodcock J. Accelerating identification and regulatory approval of investigational cancer drugs. JAMA 2011;306(23):2608−9.

[39] Khoury MJ, Bowen MS, Burke W, Coates RJ, Dowling NF, Evans JP, et al. Current priorities for public health practice in addressing the role of human genomics in improving population health. Am J Prev Med 2011;40(4):486−93.

[40] Stark Z, Dolman L, Manolio T, Ozenberger B, Hill SL, Caufield MJ, et al. Integrating genomics into healthcare: a global responsibility. Am J Hum Genet 2019;104(1):13−20.

[41] Ginsburg GS, Phillips KA. Precision medicine: from science to value. Health Aff (Millwood) 2018;37(3):694−701.

[42] Miller HD. From volume to value. Health Aff (Millwood) 2009;28(5):1418−28.

[43] Montgomery R., Valuck T., Bens C., Ferris A. Personalized medicine and quality measurement: from conflict to alignment. 2019. Retrieved from Health Affairs Forefront: healthaffairs.org/do/10.1377/forefront.20190424.431063.

[44] Mullangi S, Schleicher SM, Parikh RV. The oncology care model at 5 years - value-based payment in the precision medicine era. JAMA Oncol 2021;7(9):1283−4.

[45] Koleva-Kolarova R, Buchanan J, Vellekoop H, Huygens S, Versteegh M, Rutten-van Molken MThe HEcoPerMed Consortium. Financing and reimbursement models for personlised medicine: a systematic review to identify current models and future options. Appl Health Econ Health Policy 2022;20(4):501−24.

[46] Wholley D. Public-private partnerships in biomarker research. In: Ginsberg GS, Willard HF, editors. Genomic and personalized medicine. Elsevier; 2013. p. 474−83.

[47] Altar CA. The biomarkers consortium: on the critical path of drug discovery. Clin Pharmacol Ther 2008;83:361−4.

[48] GAIN Collaborative Research Group. New models of collaboration in genome-wide association studies. Nat Genet 2007;39:1045−51.

[49] Woodcock J. Precompetitive research: a new prescription for drug development? Clin Pharmacol Ther 2010;87:521−3.

[50] Greenstein J. Juvenile diabetes research foundation: seeding novel insights in diabetes research. Stem Cell Transl Med 2012;1(3):175−6.

CHAPTER 2

Precision approach in the medical and surgical management of newly diagnosed and refractory epilepsy

Subhan Khan[1,2,3] and Farhan A. Mirza[1,2]
[1]UK Comprehensive Epilepsy Program, University of Kentucky, Lexington, KY, United States
[2]Department of Neurosurgery, Kentucky Neuroscience Institute (KNI), University of Kentucky, Lexington, KY, United States
[3]Westchester Medical Centre, Valhalla, NY, United States

Introduction

Precision medicine is often limited to the perception of molecular or cellular therapy for human disease. In the true sense, precision medicine is any medical or surgical therapy tailored to an individual's unique disease manifestation. The disease label might be the same for a large group of individuals, but the manifestations of the disease and its impact on their life are unique. Hence, the ensuing treatment options must be tailored to their situation. Epilepsy, particularly the treatment of refractory epilepsy, fits this bill perfectly.

In the following chapter, we will briefly revisit what "epilepsy" and "precision medicine" mean. Once we have understood this foundation, we will examine how epilepsy may be approached both medically and surgically through the lens of precision medicine. This will involve novel diagnostic approaches, targeted therapy based on a genetic understanding of epilepsy, and refined surgical options based on superior localization and network assessment. This chapter will also examine the impact artificial intelligence and other burgeoning software may integrate into precision healthcare.

What is precision medicine?

Koning et al. suggest that precision medicine is not an end point but rather it is a process. This process involves continuous collection and screening of data to stratify patients into groups [1]. These groups may in turn be

The New Era of Precision Medicine
DOI: https://doi.org/10.1016/B978-0-443-13963-5.00005-4
© 2024 Elsevier Inc.
All rights reserved.

examined, resulting in feedback loops of data collection and mining resulting in continuous stratification of groups. This process aids in the creation of subpopulations that can then be examined further on a personal level.

Once we have subsets of data to examine, we can use a "bottom-to-top" approach for precision medicine as opposed to the "top-to-bottom" approach in traditional evidence-based medicine [2]. The simplest way to understand this is that the traditional approach identifies a disease or disorder and then deduces the pathogenic cause of it. This approach is based on germ theory, i.e., a patient is diagnosed with meningitis (the effect), and we can deduce based on a cause-effect relationship that they had probably been infected with a bacterium (the cause). This approach works for infectious, toxic, and Mendelian genetic diseases. However, for more complex chronic disorders such as epilepsy that are often multifactorial and with variable clinical courses, it has limited utility.

Hence, a "bottom-to-top" approach is used in complex chronic diseases such as epilepsy where the causes are examined first, e.g., multiple genetic variants (as opposed to highly pathogenic variants in a single gene as with Mendelian genetics), environmental factors, and the patient's lifestyle. Once the cause is determined, we can deduce a disorder that may arise from this. If this disorder is left unchecked, it ultimately results in the manifestation of the root problem as the disease itself. The key difference is that if precision medicine is utilized correctly, we can identify and treat the root cause before the patient contracts the disease. While this is the ideal scenario, precision medicine has not reached this level of sophistication in the field of epilepsy just yet. As a result, in the context of epilepsy, precision medicine will also be used to describe the use of novel diagnostic and therapeutic approaches to surgically treat refractory epilepsy as precisely as possible in a very literal sense.

What is a seizure?

A seizure is defined as "a transient occurrence of signs and/or symptoms due to abnormal or excessive synchronous neural activity in the brain." Epilepsy is characterized by an enduring predisposition to epileptic seizures, and it can be due to any number of causes, including vascular issues, trauma, metabolic and genetic abnormalities, abnormal brain development, migrational disorders, autoimmune conditions, neoplasia, and infections [3]. Globally, seizures affect around 10% and cause epilepsy in 1%—2% of the world's population [4]. Epilepsy may be classified into common variants

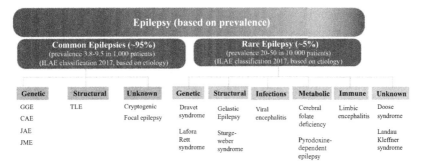

Figure 2.1 Classification of epilepsy based on its prevalence. Classification of epilepsy based on the prevalence of the disease in the global population, into common and rare epilepsies. Common epilepsy is more prevalent in people with epilepsy (~95%), and around ~5% people with epilepsy suffer from rare epilepsy syndromes. According to the latest ILAE classification in 2017, epilepsy is classified based on etiology into genetic, structural, infectious, metabolic, immune, and unknown. Different types of epilepsy are represented under each of these subcategories. *CAE*, Childhood absence epilepsy; *GGE*, genetic generalized epilepsy; *ILAE*, International League against Epilepsy; *JAE*, juvenile absence epilepsy; *JME*, juvenile myoclonic epilepsy; *TLE*, temporal lobe epilepsy.

(95% of total cases, including generalized epilepsy, absence epilepsy, temporal lobe epilepsy, etc.) and rare variants (5% of total cases, including syndromic causes such as Dravet and Sturge-Weber, infectious causes such as viral encephalitis, and metabolic causes such as cerebral folate deficiency among others) ([5], Fig. 2.1). With such a wide array of etiologies and clinical courses, epilepsy is clearly a highly complex and nuanced condition that cannot be approached with a "one-size-fits-all approach."

Stratification of patient characteristics

The promise of precision medicine is "delivering the right treatments, at the right time, every time, to the right person." Findings from traditional twin studies as well as clinical studies have shown that genetics is a factor in 70%–80% of all epilepsy cases with the other 20%–30% being due to an acquired cause such as strokes or tumors. Rare variants typically arise due to monogenic causes and are relatively simpler to investigate. Conversely, common variants tend to be multigenic, and hence assessing the plethora of potential genetic alleles that may contribute to the overall risk of epilepsy is a massive undertaking [5]. The matter becomes even more complex and nuanced when factors such as environmental factors and lifestyle are considered. Even though determining the complex interplay of factors resulting in epilepsy is a herculean task, the reward is overcoming the biggest problem of traditional evidence-based medicine: minimizing broad generalization among vastly different patient groups.

These generalizations occur across race, gender, and age. In this next section, we will briefly examine the impact such differences may have on our approach to epilepsy. Data in evidence-based medicine is typically not very representative of racial minorities. Furthermore, the data that does include racial minorities often paints racially diverse groups in the same broad strokes. For example, one study examining the prevalence of the HLA-B*5701 variant (a genetic variant linked to hypersensitivity reactions to abacavir) grouped participants as Africans and Asians. However, the prevalence ranged from 0% (Nigerian Yoruba) to 13.6% (Kenyan Maasai) in the African group and 0% (Japanese) to 17.6% (Indian Americans) in the Asian group [6]. These same generalizations occur when talking about epilepsy and antiseizure medications. A prominent example was an FDA alert on adverse reactions from carbamazepine as people with "ancestry across broad areas of Asia, including South Asian Indians, are more likely to have the HLA-B*1502 allele and should be screened for the allele." This alert was made due to studies that examined the prevalence of these allelic variants; however, the studies only examined 2 out of 37 countries in Asia [7,8].

Race is not the only factor that may influence our understanding of epileptic patients. Women with epilepsy require specific considerations that are largely overlooked when looking at full populations for data. Women have periodic hormonal changes, which may result in conditions such as catamenial epilepsy and seizure exacerbation around menopause. Additionally, many ASMs lower levels of contraceptives due to CYP-450 induction, and some contraceptives may even affect levels of certain ASMs, e.g., lamotrigine. Lastly, women who are pregnant are considered having high risk pregnancy due to the wide array of complications that may afflict the mother or child such as gestational hypertension, antepartum hemorrhage, spontaneous abortion, and preterm birth. Pregnancy also results in increased activity of CYP-450 enzymes, resulting in increased clearance of many ASMs. Lastly pregnant women on ASMs are three times more likely to give birth to a child with a birth defect than those not afflicted by epilepsy [9]. The variations in clinical course due to this multitude of factors require considerable consideration on an individual basis, which traditional medicine may not adequately address.

As with race and gender, age has a huge bearing on how epilepsy must be approached. Epilepsy has a bimodal distribution with the highest incidences at the extremes of age [4]. Elderly patients are more likely to suffer from acquired causes such as strokes, tumors, and trauma. Whereas children and infants are more likely to suffer from inherited or acquired causes. Attaining a complete history and examination with the very young presents its own unique challenge, which may make it a challenge to get a specific diagnosis. Hence, refined diagnostic modalities to ascertain genetic and metabolic abnormalities are invaluable to get a prompt diagnosis and start treatment [10].

The bottom-to-top approach in epilepsy

The bottom-to-top approach in epilepsy primarily involves the use of metabolic and genetic biomarkers and tests to identify a disorder. Ideally, we would like to find the disorder before its effects can result in a disease. However, even in circumstances where a bottom-to-top approach is used later in the course of disease or in forms of epilepsy that are congenital, e.g., tuberous sclerosis (TS), the information gleaned may be used to target our therapy to manage the root cause of the disease [11].

Metabolic biomarkers may include testing for amino acids, neurotransmitters, and autoantibodies. Examples of epilepsy that may be studied with this approach include limbic encephalitis, cerebral folate deficiency, and pyridoxine-dependent epilepsy. Pyridoxine-dependent epilepsy is particularly well studied as the genetic deficiency (ALDH7A1) results in decreased lysine metabolism, which elevates levels of α-amino-adipic semialdehyde (AASA). We can take advantage of this and test for the AASA to creatinine ratio in blood and urine [12]. With a greater understanding of the polygenic variants involved in epilepsy, we can use similar approaches to screen for and diagnose common and rare epilepsies (Fig. 2.2).

Genetic biomarkers have similar utility where they aid in the diagnosis of epilepsies with often discrete clinical findings. Patients with pathogenic variants of *PCDH19* and *SYNGAP1*, for example, present with clusters of focal febrile seizures with affective symptoms (particularly fear) and reflex seizures with bilateral eyelid myoclonus, respectively [12]. We may also use genetic biomarkers to identify disease before the full constellation of symptoms are apparent. For example, infants suffering from unilateral hemiclonic seizures during a febrile illness test positive for *SCNA1* > 90% of the time. In isolation, the clinical features point toward focal

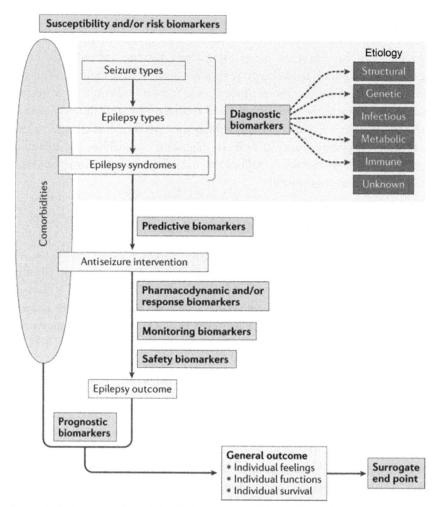

Figure 2.2 Representation of the different types of biomarkers plotted on the multilevel classification framework for the classification of epilepsy of the International League against Epilepsy [9]. Susceptibility and risk biomarkers are upstream of the classification framework. Diagnosis biomarkers provide information for the classification of epileptic syndromes and are an integral part of the epilepsy classification framework. Predictive, pharmacodynamic, response, monitoring, and safety biomarkers are used to assess the effects of antiseizure interventions. Prognosis biomarkers determine the likelihood of favorable or unfavorable outcomes in epilepsy and the associated comorbidities. A surrogate end point is generally a biomarker that is predictive of the final result of an intervention.

epilepsy, but with the genetic biomarker, we can diagnose Dravet syndrome before it fully manifests [13].

While these approaches may have great utility for rare epilepsy, which are often due to monogenic pathogenic variants, common epilepsies are trickier due to innumerous permutations and combinations of healthy and pathogenic variants that may exist in polygenic disease. As a result, the tools we need must be able to scope the overall landscape of genetic abnormalities in epilepsy. Three prominent tools for this endeavor are chromosome microarray analysis (CMA), next-generation exome sequencing (NGS), and whole genome/exome sequencing (WGS/WES).

CMA is essentially a higher magnified version of karyotyping. While traditional karyotyping can identify abnormalities that are greater than 7−10 million bases, CMA can detect changes as small as 50−100 kilobases [14]. This higher magnification lends itself to greater sensitivity as shown in the study by Sagi-Dain et al., which pitted CMA against karyotyping for genetic abnormalities in pregnancy. CMA detected changes in 4.7% of all pregnancies as opposed to 2% from karyotyping [15]. However, CMA gives no information for where the abnormality is, and it fails to detect balanced rearrangements.

NGS, WES, and WGS are all methods to encode the genome. NGS studies only a portion of the exome, and on the other end of the spectrum, we have WGS that is an exhaustive study of both the coding and noncoding portions of the genome [16]. While these studies help us identify the gene affected, it may not necessarily confer information about how that pathogenic variant translates into a mutation that translates into a disease. These variations in how the gene is affected change the nature of the mutation, the manifestation of the disease, and hence how we treat the disorder. *SCNA2* (encodes voltage-gated sodium channels) is an example of how this variation in genotype can result in different phenotypes. Truncating mutations (resulting in a loss-of-function) resulted in later milder phenotypes of epilepsy; however, these phenotypes did not respond to or worsened with sodium channel blockers. On the other hand, missense mutations (resulting in a gain-of-function) had a phenotype that responded very well to a sodium channel blocker [17]. Another problem of genome sequencing is that some genetic variants may be protective instead of reductive and the multiple allelic variants that may be detected may not necessarily affect the phenotype in equal measure.

To account for the limitations of these studies, we can use polygenic risk score (PRS) that estimates an individual's genetic chance to have a certain trait or disease. The PRS weighs the effect of common and rare genetic variants to approximate the **net** result of these allelic variants [18]. What this means for epilepsy is that it can be used to demarcate epileptic patients based on severe epilepsy, patients at risk of epilepsy, and healthy individuals. PRS has shown that the burden by common risk variants is significantly higher than that of rare variants, which supports the idea that common variants are polygenic. Additionally, PRS has demonstrated that generalized epilepsy has a significantly higher contribution from common variants than focal epilepsy does [19,20]. Utilizing the PRS to provide an overview in conjunction with a multiomics approach (genetics, metabolites, and proteins) to pinpoint the problem, we can truly zone in on the root problem(s) and target our therapy to allelic variants and metabolic derangements that are particularly burdensome [21].

Precision techniques in epilepsy

Once we have homed in on the root disorder with precision diagnostic techniques, we can then focus on how to target therapy. This precision therapy in epilepsy focuses on two aspects: targeted therapy of the disorder and secondary prevention of epileptogenesis. Targeted therapy focuses on correcting the imbalances caused by metabolic and genetic derangements. Prominent examples already discussed include using pyridoxine supplementation in pyridoxine-dependent epilepsy and sodium channel blockers for missense mutations of *SCNA2*. *SCL2A1* (encoding glucose transporter type 1) deficiencies result in impaired glucose uptake by the brain; hence, they benefit from a ketogenic diet [22,23]. Targeted therapy can also help us choose which drugs are contraindicated as in *POLG* mutations. Patients with these mutations are at risk for hepatic failure with the ingestion of valproic acid; hence they are to be avoided [24].

While these curative measures are certainly helpful, the ideal of precision medicine is to cut off the disorder from the root before disease even has a chance to manifest. Gene therapy is a tool that is still in its infancy with tools such as CRISPR to modulate the genome and prevent the conditions necessary for primary epileptogenesis. A far better understanding of the genetic variants causing and preventing epilepsy is needed before primary prevention can be attempted (although several new trials have emerged studying the effects of mTOR inhibitors on brain malformations) [25]. Hence, the focus should be on secondary prevention for conditions that have a considerable lag time between the disorder and the disease. Common examples of such conditions include Sturge-Weber and TS, which are both syndromes with a huge swath of symptoms. Epilepsy occurs in 80% of all patients with Sturge-Weber, and most of that is within the first year. A recent study by Day et al. demonstrated that seizures occurred in 25% of participants receiving preventive antiseizure medicine and in 94% of participants not receiving the preventive antiseizure medicine in their first year of life [26]. A long-term prospective trial on TS showed similar results. The trial had patients taking ASMs either before the onset of seizures or after the onset. At 24 months of age, 93% of infants receiving preventive therapy were seizure-free compared with just 35% of infants receiving standard therapy [27].

As with every other aspect of precision medicine in epilepsy, these approaches have less efficacy in epilepsy caused by polygenic pathogenic alleles. Accounting for so many variables with the information we have now is not a feasible task. To compensate for this, a systems-level approach to account for the impact of both rare and common genetic variants might be the best solution (Fig. 2.3). Delahaye-Durez et al. utilized this approach to highlight a network of 320 coexpressed genetic variants for both rare and common forms of epilepsy. This gene network was shown to be consistently downregulated in human and animal studies; hence the researchers identified ASMs that might upregulate the affected genes [28]. The downside of this approach is that it does not consider if some of those downregulations may have been protective. Therefore, study of the individual allelic variants and resultant phenotypes is still necessary in the long term.

Treatment of refractory epilepsy

Although we have made considerable progress with precision medicine approaches to epilepsy, we cannot rely on biomarkers and pharmaceuticals alone. Despite our best efforts, 30%—40% of epilepsy patients become refractory to medical therapy over time. Combination therapy becomes key in this situation, in the form of more medications, dietary changes, and consideration of surgical measures such as resection, disconnection, ablation, or neuromodulation. The goal of these efforts is singular: improvement in the quality of life by either reducing seizure burden or achieving seizure freedom. Epilepsy is a chronic disease, much like asthma, cardiac disease, etc., which are often lifelong problems. The idealistic goal in epilepsy treatment

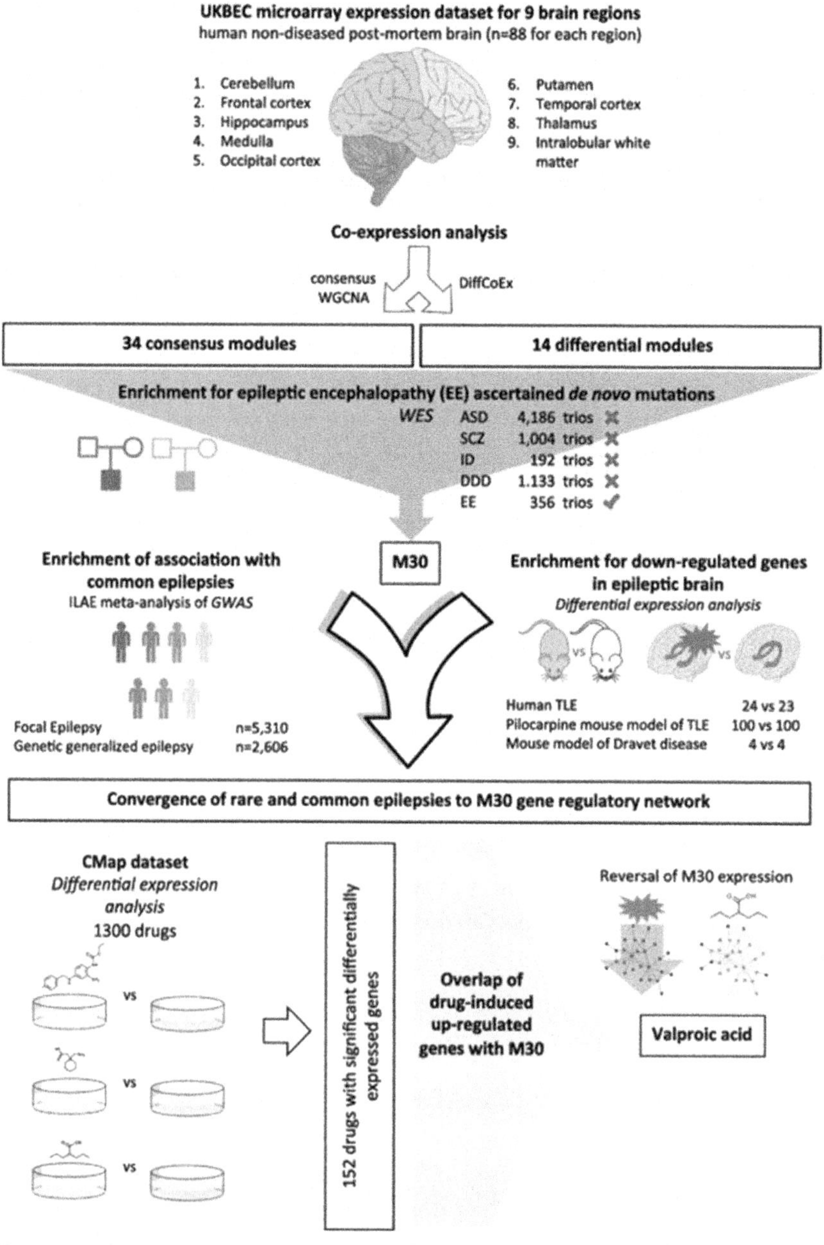

Figure 2.3 An example of the utility of a systems-level approach. Disruption of the M30 gene network leading to different types of epilepsy. M30 was tested for enrichment of association with the common forms of epilepsy using GWAS data from the ILAE metaanalysis. Analysis of network genes' expression in disease in three epilepsies suggested functional disruption and/or downregulation of the network as a common mechanism regulating susceptibility to epilepsy broadly and therefore that the network itself might be targeted as a novel therapeutic strategy. *GWAS*, Genome wide associated studies; *ILAE*, International League against Epilepsy.

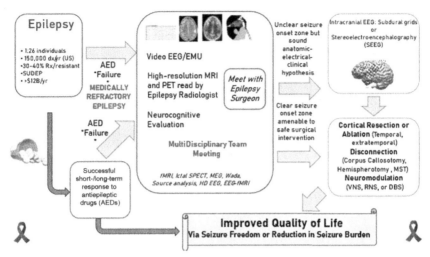

Figure 2.4 An overview of the diagnostic protocol and treatment options for refractory epilepsy. AED failure either in the short or long term necessitates further investigation for determining the seizure onset zone. Subsequently one of three modalities of surgery is performed: resection, disconnection, or modulation as monotherapy or in combination. *AED*, Antiepileptic drug; *DBS*, deep brain stimulator; *EEG*, electroencephalogram; *EMU*, epilepsy monitoring unit; *fMRI*, functional magnetic resonance imaging; *MST*, multiple subpial transection; *RNS*, responsive nerve stimulator; *SPECT*, single-photon emission computed tomography; *SUDEP*, sudden unexpected death in epilepsy; *VNS*, vagus nerve stimulator.

is seizure freedom; however, this may be an elusive target in many refractory patients. The more realistic goal for a vast majority of patients is a significant reduction in seizure burden (Figs. 2.4 and 2.5).

Unfortunately, several factors have led to continued underutilization of proven surgical methods for the treatment of refractory epilepsy. A detailed discussion of these factors is beyond the scope of this chapter, but it is important to highlight that a large part of this underutilization continues to be tethered to misinformation, inadequate referral patterns, and inadequate testing measures, which can improve a patient's candidacy for higher levels of intervention such as epilepsy surgery.

Thinking beyond the temporal lobe

Whenever epilepsy surgery is mentioned, it is almost always considered synonymous with "temporal lobectomy" for temporal lobe epilepsy. Temporal lobe epilepsy (TLE) is indeed the most common form of epilepsy.

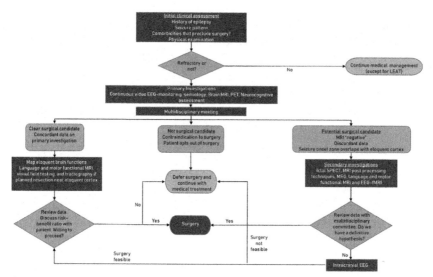

Figure 2.5 An overview of the pathways of the approach to selecting appropriate intervention for refractory epilepsy. Once epilepsy has been determined to be refractory, primary investigations are initiated and the information is reviewed by a multidisciplinary committee to evaluate whether surgical intervention is warranted. For unclear (potential) surgical candidates, secondary investigations and possible invasive monitoring (intracranial EEG) are conducted to further guide the decision-making process. *EEG*, Electroencephalography; *LAET*, long-term epilepsy-associated tumors; *MEG*, magnetoencephalography; *MRI*, magnetic resonance imaging; *PET*, positron emission tomography.

This makes sense because the signature of TLE is nicely picked up on electroencephalogram (EEG) and a robust diagnosis can be made; severe mesial temporal sclerosis (MTS) or hippocampal sclerosis is easily detectable on MRI; and high-quality studies have been conducted on temporal lobe epilepsy surgery with excellent outcomes [29]. Therefore, surgery for TLE is generally better accepted and appears to or is perceived to have overall better results than other types of epilepsy. However, there have been changing trends in the practice of refractory epilepsy that have led to a rise in the identification of patients with extratemporal lobe epilepsy (ETLE). This has largely been due to improved imaging techniques. ETLE is often considered synonymous with "nonlesional" or "MRI negative" epilepsy. Several studies have shown overall less favorable long-term results with surgery for ETLE. Two main reasons for this are inaccurate localization leading to incomplete resection of focus or epileptogenic zone or inadequate efforts to identify and visualize the epileptogenic substrate.

Converting "nonlesional" epilepsy to "lesional" epilepsy

We know that the best response to surgical treatment for epilepsy is often with any type of known lesional epilepsy such as hemispheric epilepsy, focal cortical dysplasia (FCD), TLE with MTS, epilepsy with cavernoma, or tumor related epilepsy. With this idea, several questions come to mind. Is all focal epilepsy lesional? Is all epilepsy lesional? Or in other words, is there always an epileptogenic substrate either at the cellular level or a focal anatomic substrate? If this is the case, is it simply our inability to identify that substrate, which consequently has a direct bearing on what is resected or ablated and the outcome of that resection? What if we can convert these so-called "nonlesional" cases into lesional cases with appropriate high-quality imaging and data interpretation?

The eye doesn't see what the mind doesn't know. If one is only looking for MTS or hippocampal sclerosis, one will not look for occult findings such as FCD, focal polymicrogyria, base of sulcus dysplasia, focal gyral thickening, temporal encephalocele, hypothalamic hamartoma, etc. The most common scenario is that the patient undergoes a regular brain MRI, which gets read as negative for MTS. Result: An inadequately obtained MRI is not read by an epilepsy radiologist; the patient is labeled "nonlesional" for the rest of their life, gets treated with medication or at best a vagus nerve stimulator (VNS), and a higher level of diagnostic or therapeutic surgery is never even considered. The appropriate method would be that patient undergoes a high-quality 3 Tesla brain MRI with special epilepsy sequences, which is then read by a dedicated epilepsy radiologist. Similarly, nuclear imaging was conducted and read by a dedicated epilepsy nuclear radiologist. Findings are then discussed in the setting of a refractory epilepsy conference, where all pieces of the puzzle are put into context, and in those labeled "MRI negative" previously, lesions are often discovered! Identifying a lesion completely changes the outlook for that patient. Previously considered "nonlesional" generalized epilepsy suddenly becomes a surgically amenable problem. As complexity increases tremendously in this process, this requires a whole team of experts in different areas: epileptologist, epilepsy surgeon, epilepsy radiologist, neuropsychologist, epilepsy, epilepsy nurses and coordinators, and dedicated epilepsy research team [30].

Advances in imaging methods

Several innovations have been made in diagnostic imaging modalities. These modalities include refinements in the use of MRIs, functional

MRIs (fMRIs), positron emission tomography (PET), and ictal single photon emission computer topography (SPECT) scans, as well as the advent of relatively newer modalities such as magnetoencephalograms (MEGs), high-density EEG, EEG-fMRI, resting state-fMRI, etc. [31,32]. The ability to delineate the epileptogenic zone and the associated network can lend a great deal of precision and can completely alter our approach to surgery for epilepsy.

MRIs usually operate at a strength of 1.5T (standard) and 3T (used for epilepsy protocols). Advancement in imaging technology has allowed acquisition of 7T images. In addition, modifications to these images such as MR Spectroscopy can help further isolate obscure lesions. In one study, the use of 7T MRIs and MRS in nonlesional epilepsy (from traditional images) identified focal lesions in one third of all cases [33]. Improvements in how we postprocess images has also helped us identify more focal epilepsies. A prominent example is morphometric analysis program (MAP). It is a voxel-based study that compares patient data on images to a normative database and identifies abnormalities by detecting differences such as blurring of the gray-white matter junction and abnormal extension of gray matter into white matter. MAP is very sensitive at detecting FCD when used in concordance with other diagnostic modalities such as EEG and FDG−PET and has a standalone specificity of 96% [34].

fMRIs as the name suggests greatly aid our ability to map functional areas of the brain, i.e., eloquent cortex. Although it has limited utility for mapping motor function (when compared to electrocortical stimulation or awake surgeries), diffusion tensor imaging (DTI) provides a great non-invasive method to map out white matter tracts. fMRIs are also useful for lateralizing episodic memory to the right or left temporal lobe, making them very useful for assessment in temporal lobe epilepsies [32]. fMRIs are particularly useful when mapping language functions and have been shown to be highly concordant (up to 90%) with Wada testing and neuropsychological testing [32,35,36].

Nuclear studies such as PET and ictal-SPECT have been used to localize the epileptogenic onset zone for a few years now. However, our approach to how we utilize the results has changed in recent years. PET studies excel at showing hypometabolism in temporal and extratemporal epilepsies, but the extent of hypometabolism can be beyond the epileptogenic focus. Ictal-SPECT on the other hand is mostly used in extratemporal MRI negative cases. Refinements in ictal-SPECT include subtracting interictal SPECT from ictal SPECT with MRI registration

(SISCOM) [32]. This can be made even more sensitive and specific with the use of statistical software, e.g., STATISCOM, to localize the onset zone. Although neither study truly isolates the problem, when used in conjunction with scalp EEGs, imaging studies, and neurocognitive assessment, we can narrow down our search for subsequent invasive EEG studies, and generally concordance with these studies is indicative of good surgical outcomes [37,38].

Finally, to aid in seizure localization, several new noninvasive methods have been introduced, including high-density EEG (HD-EEG) and MEG [39]. MEG is particularly helpful in identifying the seizure onset zone or focus [40]. The major downsides of MEG studies are that they are highly specialized and require dedicated facilities and personnel specializing in MEG analysis, all of which can be exorbitantly expensive.

Precision surgery in epilepsy

Localization efforts focus on pinpointing the epileptogenic zone (correlates with ictal findings), the irritative zone (correlates with interictal findings), and the functional deficit zone (correlates with examination and functional imaging). Localizing the EZ can be divided into two phases: I (EEG and imaging) (Fig. 2.6) and II (invasive monitoring).

Phase I Investigations		
Interictal state	Ictal State	Post-ictal
Physical examination Video EEG	Physical examination during ictal state Video EEG	Physical examination after seizure Video EEG
MRI PET	Ictal SPECT ASL	ASL
Neurocognitive assessment		
HD-EEG MEG		

Figure 2.6 Principles of Phase I results organization. Data collected with the different (clinical, electrophysiologic, and imaging) techniques classed in separate "channels" depending on the state of epilepsy activity they are documenting. *MEG*, Magnetoencephalography; *MRI*, magnetic resonance imaging; *PET*, positron emission tomography; *SPECT*, single-photon emission computed tomography.

Phase II is indicated in three major scenarios:

1. Data from Phase I is not concordant for localizing the EZ or
2. Data from Phase I shows that the EZ is sitting in or near eloquent cortex; hence Phase II studies are used to mark borders for a minimalist approach to surgery [32,41].

The two major modalities are subdural grid and depth electrodes (SDG/DE) and stereoelectroencephalography (SEEG). Once electrodes are placed, the patient is monitored via EEG, and the electrodes may be stimulated to induce seizures. This approach allows us to get a more detailed look at which areas correspond to ictal and interictal changes with a greater degree of reliability.

In Europe, for decades, the SEEG methodology has been the mainstay for Phase II monitoring for refractory epilepsy. SEEG has become popular in the United States in the last two decades. Compared to traditional subdural grids, SEEG provides a more refined understanding of the seizure focus and its associated network. With SEEG, the depths of the brain can be explored without having to perform a large craniotomy and implant large grids. The spatial and temporal resolution of information obtained during SEEG analysis is akin to a 3D map of the epileptogenic network, allowing a more nuanced understanding. Its only real limitation is that it is a very targeted approach; hence the sampling sites must be chosen very carefully; otherwise the EZ and lesion can be missed completely. Therefore, a robust preimplantation hypothesis is imperative for a successful implantation. This can be achieved only if each individual patient's epilepsy is studied rigorously with the aforementioned imaging methods, concordance or discordance of data is determined, and then a solid hypothesis is developed before embarking on SEEG [42].

Intracranial monitoring in general has shown to be highly effective in localizing the EZ when other modalities have revealed discordant data [43]. Several studies have shown that the EZ was identified in up to 92% of all cases tested when Phase I studies failed to localize the lesion [44,45]. Invasive studies also excel at detecting subclinical seizures (SCS) in patients with refractory focal epilepsy with one study detecting SCS in two-thirds of the total patient cohort [46].

The aim of precision in epilepsy surgery is to identify the epileptogenic substrate or network and then demarcate a clear area of resection, since resections often have the best outcomes. Disconnection strategies are also useful, especially in multilobar epilepsies. Neuromodulation in the form of vagus nerve stimulation (VNS), deep brain stimulation (DBS), and now responsive neurostimulation (RNS) has altogether changed the landscape for surgical treatment of refractory epilepsy. These methodologies have opened avenues for patients who were previously considered poor candidates for surgical intervention. RNS is a remarkable treatment method based on a "closed loop" system, which automatically monitors electric activity in the brain, predicts aberrant activity, and provides a current to abort seizure activity as proactively as possible. It also provides more chronic local electrocorticography for the clinician to guide further steps in management [47]. Additionally, RNS therapy has improved treatment of refractory epilepsy patients who have a wide-area network or a thalamic generator. Intracranial recording from thalamic targets and their correlation with focal EEG findings have ushered in new methods of targeting the thalamus and the active node in the network. This has led to RNS being implanted in the pulvinar, anterior nucleus of thalamus (ANT), or centromedian (CM) nuclei with remarkable results [48,49].

Nevertheless, these measures are often alternatives to resection, which is why the focus in precision medicine has been on innovative means to conduct resective or ablative surgery. Resection of the epileptogenic focus is the gold standard for the treatment of refractory epilepsy. To achieve the same results without open surgery, several approaches have been studied and attempted, such as stereotactic radiosurgery, MR-guided laser interstitial thermal therapy, MR-guided focused ultrasonography, and SEEG-guided radiofrequency thermocoagulation. The method is conceptually the same with all modalities causing focused tissue damage [42]. Results are encouraging but still not as favorable as resective strategies. Regardless of which treatment strategy is chosen, treatment is tailored to each individual's disease, making it a true example of precision medicine (Fig. 2.7).

Future implications and conclusion

Clearly, precision treatment in epilepsy is extremely dense and multifactorial. The sheer breadth and depth of the information needed to conduct precision medicine in epilepsy warrants the use of a robust multidisciplinary approach. Specialists in epilepsy, epilepsy surgery, neuropsychology, neuroradiology, and epilepsy research all need to work in tandem to tailor treatment to each patient's needs and guide individualized treatment. To aid in these endeavors, artificial intelligence and integrative software are now being developed to streamline data flow among the different stages of epilepsy surgery and among different specialists to provide smooth and

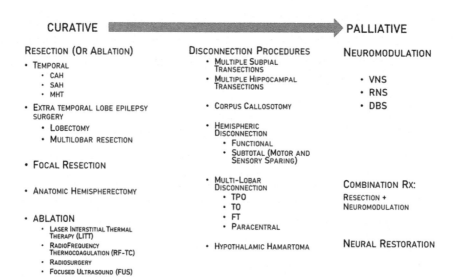

Figure 2.7 Different modalities for surgical intervention in epilepsy. *CAH,* cortico-amygdalohippocampectomy; *DBS,* Deep brain stimulator; *FT,* frontotemporal; *MHT,* multiple hippocampal transection; *RNS,* responsive nerve stimulator; *SAH,* selective amygdalohippocampectomy; *TO,* temporo-occipital; *TPO,* temporoparietooccipital; *VNS,* vagus nerve stimulator.

integrated decision-making [50]. Similarly artificial intelligence can utilize computational machines with specific functions, such as automatic diagnosis, prognosis prediction, and seizure onset localization for each patient using machine learning techniques based on large normative databases. Furthermore, by recreating the brain network dynamics of a particular patient based on the individual patient's data, we can weigh multiple modalities of information to target our approach even further [51]. Although much work remains to be done, the vast uncharted territories for precision medicine in epilepsy provide a glimpse of a cautiously optimistic future for clinicians and patients alike.

References

[1] König IR, Fuchs O, Hansen G, von Mutius E, Kopp MV. What is precision medicine? Eur Respiratory J 2017;50(4):1700391.
[2] Whitcomb DC. Primer on precision medicine for complex chronic disorders. Clin Transl Gastroenterol 2019;10(7):e00067.
[3] Johnson EL. Seizures and epilepsy. Med Clin North Am 2019;103(2):309−24.
[4] Beghi E. The epidemiology of epilepsy. Neuroepidemiology. 2020;54(2):185−91.

[5] Thakran S, Guin D, Singh P, Singh P, Kukal S, Rawat C, et al. Genetic landscape of common epilepsies: advancing towards precision in treatment. Int J Mol Sci 2020;21(20).

[6] Rotimi CN, Jorde LB. Ancestry and disease in the age of genomic medicine. N Engl J Med 2010;363(16):1551−8.

[7] Payne PW. Ancestry-based pharmacogenomics, adverse reactions and carbamazepine: is the FDA warning correct? Pharmacogenomics J 2014;14(5):473−80.

[8] Callier SL. The use of racial categories in precision medicine research. Ethn Dis 2019;29(Suppl 3):651−8.

[9] Li Y, Zhang S, Snyder MP, Meador KJ. Precision medicine in women with epilepsy: The challenge, systematic review, and future direction. Epilepsy Behav 2021; 118:107928.

[10] Sharma P, Hussain A, Greenwood R. Precision in pediatric epilepsy. F1000Res 2019;8.

[11] Scheffer IE, Berkovic S, Capovilla G, Connolly MB, French J, Guilhoto L, et al. ILAE classification of the epilepsies: position paper of the ILAE commission for classification and terminology. Epilepsia. 2017;58(4):512−21.

[12] Nabbout R, Kuchenbuch M. Impact of predictive, preventive and precision medicine strategies in epilepsy. Nat Rev Neurol 2020;16(12):674−88.

[13] Hattori J, Ouchida M, Ono J, Miyake S, Maniwa S, Mimaki N, et al. A screening test for the prediction of Dravet syndrome before one year of age. Epilepsia. 2008; 49(4):626−33.

[14] Levy B, Wapner R. Prenatal diagnosis by chromosomal microarray analysis. Fertil Steril 2018;109(2):201−12.

[15] Sagi-Dain L, Maya I, Reches A, Frumkin A, Grinshpun-Cohen J, Segel R, et al. Chromosomal microarray analysis results from pregnancies with various ultrasonographic anomalies. Obstet Gynecol 2018;132(6):1368−75.

[16] A roadmap for precision medicine in the epilepsies. Lancet Neurol 2015; 14(12):1219−28.

[17] Abbott CM. Precision medicine in epilepsy—the way forward? ACS Chem Neurosci 2019;10(4):2080−1.

[18] Takata A, Nakashima M, Saitsu H, Mizuguchi T, Mitsuhashi S, Takahashi Y, et al. Comprehensive analysis of coding variants highlights genetic complexity in developmental and epileptic encephalopathy. Nat Commun 2019;10(1):2506.

[19] Speed D, O'Brien TJ, Palotie A, Shkura K, Marson AG, Balding DJ, et al. Describing the genetic architecture of epilepsy through heritability analysis. Brain. 2014;137(Pt 10):2680−9.

[20] Leu C, Stevelink R, Smith AW, Goleva SB, Kanai M, Ferguson L, et al. Polygenic burden in focal and generalized epilepsies. Brain. 2019;142(11):3473−81.

[21] Striano P, Minassian BA. From genetic testing to precision medicine in epilepsy. Neurotherapeutics. 2020;17(2):609−15.

[22] Neal EG, Chaffe H, Schwartz RH, Lawson MS, Edwards N, Fitzsimmons G, et al. The ketogenic diet for the treatment of childhood epilepsy: a randomised controlled trial. Lancet Neurol 2008;7(6):500−6.

[23] Reif PS, Tsai M-H, Helbig I, Rosenow F, Klein KM. Precision medicine in genetic epilepsies: break of dawn? Expert Rev Neurotherapeutics 2017;17(4):381−92.

[24] Engelsen BA, Tzoulis C, Karlsen B, Lillebø A, Laegreid LM, Aasly J, et al. POLG1 mutations cause a syndromic epilepsy with occipital lobe predilection. Brain. 2008;131(Pt 3):818−28.

[25] Demarest S, Brooks-Kayal A. Precision treatments in epilepsy. Neurotherapeutics. 2021;18(3):1429−31.

[26] Day AM, Hammill AM, Juhász C, Pinto AL, Roach ES, McCulloch CE, et al. Hypothesis: presymptomatic treatment of sturge-weber syndrome with aspirin and antiepileptic drugs may delay seizure onset. Pediatr Neurol 2019;90:8—12.

[27] Jozwiak S, Słowińska M, Borkowska J, Sadowski K, èojszczyk B, Domańska-Pakieła D, et al. Preventive antiepileptic treatment in tuberous sclerosis complex: a long-term, prospective trial. Pediatr Neurol 2019;101:18—25.

[28] Delahaye-Duriez A, Srivastava P, Shkura K, Langley SR, Laaniste L, Moreno-Moral A, et al. Rare and common epilepsies converge on a shared gene regulatory network providing opportunities for novel antiepileptic drug discovery. Genome Biol 2016; 17(1):245.

[29] Wiebe S, Blume WT, Girvin JP, Eliasziw M. A randomized, controlled trial of surgery for temporal-lobe epilepsy. N Engl J Med 2001;345(5):311—18.

[30] Duncan JS. Selecting patients for epilepsy surgery: synthesis of data. Epilepsy Behav 2011;20(2):230—2.

[31] Zijlmans M, Zweiphenning W, van Klink N. Changing concepts in presurgical assessment for epilepsy surgery. Nat Rev Neurol 2019;15(10):594—606.

[32] Duncan JS, Winston GP, Koepp MJ, Ourselin S. Brain imaging in the assessment for epilepsy surgery. Lancet Neurol 2016;15(4):420—33.

[33] Sidhu MK, Duncan JS, Sander JW. Neuroimaging in epilepsy. Curr Opin Neurol 2018;31(4):371—8.

[34] Wong-Kisiel LC, Tovar Quiroga DF, Kenney-Jung DL, Witte RJ, Santana-Almansa A, Worrell GA, et al. Morphometric analysis on T1-weighted MRI complements visual MRI review in focal cortical dysplasia. Epilepsy Res 2018;140:184—91.

[35] Bohm P, McKay J, Lucas J, Sabsevitz D, Feyissa AM, Ritaccio T, et al. Wada testing and fMRI in a polyglot evaluated for epilepsy surgery. Epileptic Disord 2020; 22(2):207—13.

[36] Jeong JW, Asano E, Juhász C, Chugani HT. Quantification of primary motor pathways using diffusion MRI tractography and its application to predict postoperative motor deficits in children with focal epilepsy. Hum Brain Mapp 2014; 35(7):3216—26.

[37] Peedicail JS, Singh S, Molnar CP, Numerow LM, Gnanakumar R, Josephson CB, et al. Impact of ictal subtraction SPECT and PET in presurgical evaluation. Acta Neurol Scand 2021;143(3):271—80.

[38] Juhász C, John F. Utility of MRI, PET, and ictal SPECT in presurgical evaluation of non-lesional pediatric epilepsy. Seizure. 2020;77:15—28.

[39] Plummer C, Vogrin SJ, Woods WP, Murphy MA, Cook MJ, Liley DTJ. Interictal and ictal source localization for epilepsy surgery using high-density EEG with MEG: a prospective long-term study. Brain. 2019;142(4):932—51.

[40] El Tahry R, Wang ZI, Thandar A, Podkorytova I, Krishnan B, Tousseyn S, et al. Magnetoencephalography and ictal SPECT in patients with failed epilepsy surgery. Clin Neurophysiol 2018;129(8):1651—7.

[41] Chauvel P, Gonzalez-Martinez J, Bulacio J. Chapter 3 — Presurgical intracranial investigations in epilepsy surgery. In: Levin KH, Chauvel P, editors. Handbook of clinical neurology, 161. Elsevier; 2019, p. 45—71.

[42] Baumgartner C, Koren JP, Britto-Arias M, Zoche L, Pirker S. Presurgical epilepsy evaluation and epilepsy surgery. F1000Res 2019;8.

[43] Rosenow F, Menzler K. Invasive EEG studies in tumor-related epilepsy: when are they indicated and with what kind of electrodes? Epilepsia. 2013;54(s9):61—5.

[44] Dzhafarov VM, Guzeeva AS, Amelina EV, Khalepa AA, Dmitriev AB, Denisova NP, et al. [Invasive EEG for temporal lobe epilepsy: selection of technique]. Zh Vopr Neirokhir Im N N Burdenko 2021;85(5):23—9.

[45] Méreaux JL, Gilard V, Le Goff F, Chastan N, Magne N, Gerardin E, et al. Practice of stereoelectroencephalography (sEEG) in drug-resistant epilepsy: Retrospective series with surgery and thermocoagulation outcomes. Neurochirurgie. 2020;66(3):139—43.

[46] Tsuboyama M, Harini C, Liu S, Zhang B, Bolton J. Subclinical seizures detected on intracranial EEG: Patient characteristics and impact on surgical outcome in a single pediatric epilepsy surgery center. Epilepsy Behav 2021;121(Pt A):108040.

[47] Hoffman CE, Parker WE, Rapoport BI, Zhao M, Ma H, Schwartz TH. Innovations in the Neurosurgical Management of Epilepsy. World Neurosurg 2020;139:775—88.

[48] Kwon CS, Schupper AJ, Fields MC, Marcuse LV, La Vega-Talbott M, Panov F, et al. Centromedian thalamic responsive neurostimulation for Lennox-Gastaut epilepsy and autism. Ann Clin Transl Neurol 2020;7(10):2035—40.

[49] Sisterson ND, Kokkinos V, Urban A, Li N, Richardson RM. Responsive neurostimulation of the thalamus improves seizure control in idiopathic generalised epilepsy: initial case series. J Neurol Neurosurg Psychiatry 2022;93(5):491—8.

[50] Higueras-Esteban A, Delgado-Martínez I, Serrano L, Principe A, Pérez Enriquez C, González Ballester M, et al. SYLVIUS: a multimodal and multidisciplinary platform for epilepsy surgery. Comput Methods Prog Biomed 2021;203:106042.

[51] An S, Kang C, Lee HW. Artificial intelligence and computational approaches for epilepsy. J Epilepsy Res 2020;10(1):8—17.

CHAPTER 3

In vitro diagnostics in oncology in precision medicine ☆

Donna M. Roscoe

Division of Molecular Genetics and Pathology, Office of In Vitro Diagnostics, Office of Product Evaluation and Quality Center, Center for Devices and Radiological Health, U.S. Food and Drug Administration, Silver Spring, MD, United States

Precision medicine is intended to transform health care by customizing treatment plans to each patient's unique "genes, environment and lifestyle" [1]. The era of precision medicine is made possible by the scientific and technological advances that enable exploring the role of unique genomic characteristics in specimens from patients with the disease, enabling significant advances for patient care, and providing more information and options to health care providers to improve patient management as evidenced by the expansive number of biomarker-based therapeutics [2]. Successful implementation of precision medicine is therefore dependent on highly accurate and reliable diagnostic tests. The FDA (US Food and Drug Administration) is instrumental in supporting the success of precision medicine, with numerous leading-edge initiatives to support the radical sea change that is occurring in drug and diagnostic development. This chapter provides an overview of regulatory review to ensure high-quality, safe, and effective diagnostics and discusses regulatory strategy and initiatives for developing increasingly sophisticated diagnostics in precision medicine.

Regulation of in vitro diagnostics

FDA regulates medical devices, which include in vitro diagnostics (IVDs). The Center for Devices and Radiological Health (CDRH) is responsible for premarket review and postmarket surveillance of most IVDs. IVDs are defined as "those reagents, instruments, and systems intended for use in the diagnosis of disease or other conditions, including a determination of

---

☆ This chapter represents the views of the author and should not be considered formal or binding FDA guidance or policy.

The New Era of Precision Medicine
DOI: https://doi.org/10.1016/B978-0-443-13963-5.00003-0

© 2024 Elsevier Inc.
All rights reserved.

Table 3.1 Regulatory classes and controls.

Classification	Risk	Regulatory controls	Premarket Submission type	Examples
Class I	Low to moderate risk	General controls	Exempt[a]	Select Immunohistochemistry antibodies
Class II	Moderate	General controls and special controls	510(k)[b] De Novo	Tests for monitoring, prognosis, and genetic health risk
Class III	High	General controls and premarket approval	PMA	Tests for cancer screening

PMA, Premarket approval.
[a]While most Class I devices are exempt from premarket review, some are subject to 510(k) premarket notification.
[b]While most Class II devices are subject to 510(k) premarket notification, some are exempt from premarket submission.

the state of health, in order to cure, mitigate, treat, or prevent disease or its sequelae. Such products are intended for use in the collection, preparation, and examination of specimens taken from the human body" [3]. The Federal Food, Drug, and Cosmetic Act, section 513 establishes a risk-based classification system for medical devices[1] that includes regulatory controls needed to provide a reasonable assurance of safety and effectiveness of the device (see Table 3.1).

For IVDs determination of risk is generally based on the potential for harm to a patient if the test does not work as expected. Different levels of regulatory controls are used to mitigate the risks. Regulatory controls may include general controls, special controls, or premarket approval (PMA) application and are used to mitigate the risks for low-, moderate-, and high-risk devices, respectively.

General controls apply to all IVDs with rare exceptions. Examples of general controls include, but are not limited to, registration of the companies and listing of their products with the FDA, record-keeping

[1] The terms device, IVD, and test are used interchangeably in this document.

requirements, adverse event reporting, and quality system regulation (QSR) including good manufacturing practice (GMP) requirements. Low-risk, Class I devices are generally only required to comply with general control requirements subject to certain limitations. An example of a low-risk diagnostic test is a Follicle-Stimulating Hormone Test System [4].

Moderate-risk Class II devices are subject to special controls in addition to general controls. Special controls are regulatory requirements for Class II devices, which are devices for which general controls alone are insufficient to provide reasonable assurance of the safety and effectiveness of the device, and for which there is sufficient information to establish special controls to provide such assurance. "Special controls" does not refer to the use of reagents in an assay, but rather the regulatory processes that may be applicable to mitigating the risks of wrong results. Special controls for diagnostic devices usually include a premarket review by the FDA of a submission referred to as a premarket notification or "510(k)" (so-called after the section in the Federal Food, Drug, and Cosmetic (FD&C) Act. Sponsors submitting 510(k)s for IVDs are required to generate the data outlined in the regulation for that device type and indication for use, which include analytical validation and method comparison to a predicate device already cleared by the Agency. IVDs authorized via the 510(k) process are considered "cleared" whereas IVDs authorized via PMA are considered "approved". FDA "cleared" products have been found by the FDA to be "substantially equivalent" to a legally marketed device. For many moderate-risk oncology IVDs, clinical validation is provided in situations where technological and biological differences raise issues such that a method comparison between two tests is inadequate to demonstrate substantial equivalence. A prognostic gene signature test is an example of a test for which an applicant would provide independent clinical data supporting its use. The analytical and clinical performance are described in the device "labeling," including the instructions for use, and additionally include information about the limitations of the assay used that may include limits of the test function and validation, potential sources of error, and limits to the interpretation of results. The evaluation of test performance and the reporting of these results and limitations in the labeling are intended to mitigate the risks of incorrect results and incorrect interpretation of the results. Some moderate-risk IVDs may be 510(k) exempt provided they follow the special controls for the specific test type, as required and described in the regulation. One example of a

Class II IVD that is exempt from 510(k) notification is the "Autosomal recessive carrier screening gene mutation detection system" (21 CFR 866.5940) [5].

High-Risk Class III IVDs are those tests that are "intended to be used in supporting or sustaining human life or preventing impairment of human health, or that may present a potential unreasonable risk of illness or injury for which general controls and special controls are insufficient to provide reasonable assurance of the safety and effectiveness of a device." For this reason, most high-risk IVDs require the approval of a PMA application prior to marketing the device. PMA submissions include a thorough review of analytical and clinical validation. There is also an inspection of manufacturing facilities and a review of the manufacturer's quality system (QS) information. A PMA may include a review by FDA's Bioresearch Monitoring (BIMO) program to assess the study conduct and integrity of the data. BIMO will perform on-site inspections, data audits, and reviews of documentation such as financial disclosures submitted in support of the PMA. They will also check to determine that the rights of patients were protected and consistent with the appropriate laws in effect. This includes, for example, IRB approval and informed consent documents.

Many diagnostic tests in the oncology setting that are first-of-a-kind IVDs are by law automatically Class III, even though they may be considered low- or moderate-risk tests. These tests may be eligible for the De Novo classification process [6]. This process provides a pathway to Class I or Class II classification for medical devices for which general controls or general and special controls provide a reasonable assurance of safety and effectiveness, but for which there is no legally marketed predicate device. First-of-a-kind moderate- and low-risk devices are reviewed for safety and effectiveness and establish a path for substantial equivalence for similar devices. The FDA maintains a database of successful De Novo decisions [7]. The database includes access to the classification order that lists the special controls for all future devices of this type, and the decision summary that provides transparency regarding the data that was provided for FDA review in support of the original device.

Diagnostic test developers seeking to market tests for rare diseases may consider marketing their test as a Humanitarian Use Device (HUD), which is a medical device intended for diseases or conditions that affect fewer than 8000 individuals in the United States per year. For diagnostic tests, this number takes into consideration the total number of patients that would receive the test and not just the incidence of positive cases in the population. Receiving a HUD enables the test developer to submit a Humanitarian Device Exemption

(HDE), which is a type of marketing application for HUD products. Due to the rarity of the disease or condition and to promote access to diagnostic tests for these rare diseases, a submitter of an HDE only needs to demonstrate probable benefit and is not required to demonstrate effectiveness as described in Section 515 of the FD&C Act. However, a HUD with an approved HDE is subject to certain restrictions on profit, and IRB and informed consent requirements for use may apply [8].

Quality system regulation/good manufacturing practices

Medical devices are subject to the QS Regulation (QSR) (21 CFR part 820), including "Good Manufacturing Processes" (GMP) and cover 15 sections designed to follow the product life cycle for a device. These requirements include, but are not limited to, design controls, purchasing controls, production and process controls, acceptance activities, corrective and preventive actions, labeling and packing controls, handling, storage, distribution and installation, records management, servicing, statistical method, and audits [9,10].

Reporting of Corections and Removals

Diagnostic manufacturers are required to inform the FDA when problems with their devices are encountered (21CFR Pt 806 and 21 CFR Pt 806). Depending on the nature of the problem, FDA will work with the sponsor to achieve a successful recall and/or corrective action. The FDA also relies on the healthcare community and the public to report to FDA any issues they have with products. The reporting is voluntary but is critical to FDA maintaining and improving patient safety.[2]

Investigational diagnostics in precision medicine

A steadily increasing number of clinical trials rely on the use of tests to identify biomarkers that are intended to be predictive of response. These biomarkers may be novel genomic, protein, or omics biomarkers or may be commonly assessed in the clinical laboratory as part of a diagnostic testing panel. As noted, the FDA expects tests to be validated both

[2] U.S. Food & Drug Administration. Medical Device Reporting (MDR): How to Report Medical Device Problems, https://www.fda.gov/medical-devices/medical-device-safety/medical-device-reporting-mdr-how-report-medical-device-problems

analytically (the test is reliably measuring the analyte it is intended to measure) and clinically (the test result is clinically meaningful), and this validation is specific to the test results and the specific indication for use. Tests that are not FDA-cleared or approved, or may be cleared or approved but are not cleared or approved for the specific intended use and indications for use, are considered investigational. This includes tests that detect a biomarker for an investigational therapeutic because the safe and effective use of these tests as predictors of patients' response to the specific therapeutic is unknown.

Studies using investigational tests fall into one of three categories: exempt, nonsignificant risk, or significant risk. All clinical evaluations of investigational devices, unless exempt, must have an approved IDE before the study is initiated [11]. Exempt device studies are those where the results of the investigational tests used in the study do not put subjects at risk, for example, if results are not reported to patients and physicians. Although, these studies are exempt from requirements in 21 CFR part 812, they are not exempt from human subject protection requirements under 21 CFR Part 50, including, but not limited to, informed consent and initial and continuing institutional review board (IRB) review requirements under 21 CFR part 56. For all nonexempt studies, the designation of nonsignificant risk vs. significant risk is dependent on several factors. Significant risk devices are defined in 21 CFR 812.3(m) based on the applicability of one or more of the following:

1. Is intended as an implant and presents a potential for serious risk to the health, safety, or welfare of a subject;
2. Is purported or represented to be for use in supporting or sustaining human life and presents a potential for serious risk to the health, safety, or welfare of a subject;
3. Is for a use of substantial importance in diagnosing, curing, mitigating, or treating disease, or otherwise preventing impairment of human health and presents a potential for serious risk to the health, safety, or welfare of a subject; or
4. Otherwise presents a potential for serious risk to the health, safety, or welfare of a subject.

A nonsignificant risk device is not defined in the regulation and therefore is the designation when the investigational device is neither exempt nor significant risk. Nonsignificant risk devices are considered to have a deemed approved IDE, and the sponsor is expected to follow the abbreviated IDE requirements. In addition to requirements for informed consent and IRB approval, highlights of the abbreviated requirements stipulate that the sponsor of the investigation must appropriately label the investigational IVD, comply with the monitoring requirements, maintain appropriate records, and not promote or market the investigational IVD nor represent the investigational IVD as being safe or effective for the purposes for which it is being investigated.

When tests are used to determine eligibility for a clinical trial of a drug, poor performing tests present risks to patients, particularly cancer patients, when a test result is wrong. These risks include, but are not limited to, losing valuable time entering a trial for which there is no expectation that they may derive benefit, and using tumor tissue that may be needed for additional tests. For these reasons, clinical trial sponsors and IRBs assess the risk of the use of the tests that are determining eligibility for the investigational therapeutic trial. Patients in investigational clinical trials that are proceeding under an investigational new drug (IND) application are being monitored for safety as part of the clinical trial, and therefore, the risk considerations for the investigational test are focused specifically on the risks posed by the test. In other words, given that the patient has consented to the risks associated with the investigational therapeutic, consideration is given to the additional risks posed by the use of the test if the test does not perform as expected in the specific clinical setting. The criteria that may be helpful for this evaluation are described in the draft guidance document,[3] "Investigational IVD Used in Clinical Investigations of Therapeutic Products; Draft Guidance for Industry, Food and Drug Administration Staff, Sponsors, and Institutional Review Boards" [12]. The criteria for determination of the risks based on the use of a noninvestigational device are as follows: Will use of the results from an investigational IVD lead to some study subjects foregoing or delaying a treatment that is known to be effective? Will use of the results from an investigational IVD expose study subjects to safety risks (e.g., adverse events from the investigational therapeutic product) that exceed the risks encountered with the control arm therapy or nontrial standard of care? Is it likely, based on existing knowledge about the relationship between the biomarker and the investigational therapeutic product, that incorrect

---

[3] FDA Guidance documents are documents prepared for FDA staff, the regulated industry, and/or the public that describe the Agency's interpretation of, or policy on, a regulatory issue. Unlike statutes and regulations, guidance documents themselves cannot generally create legally binding requirements. Draft guidance is not finalized guidance; it is not for implementation as well as contains nonbinding recommendations.

results from the investigational IVD would present a potential for serious risk to study subjects? or Will the study require an invasive sampling procedure for the purpose of the test that is not part of the care in the target population that could lead to serious morbidity or mortality. While each trial is considered on a case-by-case basis, a common scenario is one in which the trial is enrolling all patients but stratifying on the biomarker result. In these cases, the investigation of the device is generally considered a nonsignificant risk because there are no additional risks posed to the patient based on the use of the test because everyone is enrolled irrespective of the test result. However, if the study were enrolling only biomarker-positive patients with newly diagnosed cancer for whom there are approved and effective therapeutic options, the use of the investigational device could be considered significant risk because the decision to enter the trial includes need for a test result with acceptable performance.

Diagnostics in personalized medicine

Medical laboratory tests are essential to the clinical management of patients. High-quality patient management begins with accurate and reliable diagnostic tests. If tests are not accurate and reliable, false results can lead to unnecessary confirmatory testing, failure to have appropriate follow-up testing, an unnecessary treatment that can be invasive or have harmful side effects, delays in receiving an accurate diagnosis leading to delays in subsequent treatments, or failure to have timely follow-up due to a false sense of security. As already noted, both analytical and clinical validation of a test ensure that the risks of wrong results are mitigated and that the test is safe and effective for the proposed use. The studies appropriate to validate the performance of a test are dependent on the intended use, indications for use, and test technology. The intended use statement describes how the device is to be used. This includes whether the test is qualitative or quantitative, specimen type (e.g., blood, saliva, swab, tumor tissue, plasma, and urine), analytes or biomarkers to be measured or detected, technology (e.g., high-throughput sequencing, qPCR, immunohistochemistry, and flow cytometry), and any specifics regarding the target population. The indication for use describes the target patient population (e.g., individuals over 50 years of age) and the clinical purpose (e.g., diagnosis, monitoring, prognosis, risk prediction, screening, early detection, or response to treatment). The validation of the device is relevant to these specifics and with consideration of the range of results.

Analytical validation

Analytical validation is critical to demonstrate that the test is robust and reliable for the proposed use. The data from these studies informs the users of the tests (such as laboratories, physicians, and the public) of the validated uses and expectations for the test. These studies are expected to be conducted with clinical samples to represent real-world use. When clinical specimens are not available due to limited volume or rarity of the biomarker, for example, then contrived specimens may be acceptable but should closely mimic the clinical specimen to the extent possible. The specimens tested should represent the range of results with particular attention to specimens near the clinical decision point. Table 3.2 provides an overview of the types of key studies and considerations given to each study. Studies applicable to novel technologies may be useful to gain a greater understanding of the device. Analytical validation includes the entire finalized test system, including all reagents, instruments and software, and the entire workflow as described in the instructions for use. Analytical validation should be performed in advance of clinical validation.

There are numerous resources and consensus standards for sponsors seeking to obtain information about study designs to support the analytical validation of their test. For example, in addition to device-specific guidance that FDA makes available online [13], the Clinical and Laboratory Standards Institute (CLSI) is a not-for-profit standards development organization that develops laboratory consensus standards intended to support medical laboratory testing. The FDA may work with other members of a CLSI committee to support the development of the study designs for test validation described in these documents. A list of FDA- recognized consensus standards is available online [14]. The FDA also makes transparent the review summary of all cleared or approved diagnostic tests, including summaries of the analytical validation studies performed. These decision summaries are publicly available and accessible via FDA databases [15] and provide useful references for the types of study designs appropriate to support the analytical test performance.

Clinical validation

Clinical validation establishes the relationship between a biomarker and a condition of interest and that the biomarker is fit for purpose. Clinical validation is generally included for De Novos, and PMAs, and for 510(k)s for devices where a method comparison to the predicate is not adequate

Table 3.2 Types of analytical performance studies.

Study[a]	Description
Accuracy	Refers to the trueness of a measurement. Accuracy is typically demonstrated either through a comparison of the test results to a reference method or a validated orthogonal method. Accuracy should not use a prototype of the device under review.
Precision	Precision refers to the closeness in measurements and includes the evaluation of replicates under variable conditions such as test users, days, runs, within-run, instruments, reagent lots, and test variables based on the test components and protocols. Although assays may be qualitative in reporting (e.g., positive vs. negative), if there is underlying quantitative measurement (e.g., cycle thresholds and variant allele frequencies), precision of these measurements may also be assessed to understand the potential for wrong results. The precision of controls and calibrators are also assessed in the precision studies to provide evidence of their fit for purpose use.
Analytical Sensitivity/LoD	Refers to the lower detection limits of the assay and includes LoB, LoD, and for quantitative assays, LoQ. LoD studies may include the analytical sensitivity of the test for the total analyte, as well as the minimal proportion of a specific analyte within the total concentration of the analyte (e.g., lowest variant allele frequency detectable).
Analytical specificity	Refers to the ability to detect the intended target and no other targets and includes the evaluation of exclusivity for assays with multiple targets, cross-reactivity to assess detection of off targets, and interference that evaluates the potential for both endogenous and exogenous sources to cause false-negative and false-positive results.
Linearity	The ability to provide quantitative test results that are directly proportional to the concentration of the measurand (quantity to be measured) in a test sample.

(Continued)

Table 3.2 (Continued)

Study[a]	Description
Carryover/crosscontamination	Evaluation of whether there is a carryover of analytes (e.g., amplification products for nucleic-based tests) across samples tested, which can lead to false-positive results.
Matrix equivalence	A demonstration that different specimen types yield clinically equivalent results.
Guard-band/robustness	Testing the IVD under varying conditions across steps where improper use or variations, such as time and temperature, may occur. These studies establish the robustness of an IVD.
Specimen collection, storage, handling, and stability	Validation of specific collection devices; stability conditions such as freeze-thaw, shipping, and storage; and processed samples, such as cut-slide stability, are important to assess how sources of variability may impact downstream testing.
Preanalytical	Preanalytical steps include specimen handling and may include preliminary isolation/extraction/purification/modification of the analyte prior to testing, e.g., nucleic acid extraction, bisulfate modification, and macro/microdissection. These steps are validated using a representative range of conditions and may be included in the guard band studies.
Reagent stability	Includes storage, shipment, freeze-thaw, and in-use/open stability studies for the reagents.
Contrived specimen functional characterization	A study designed to demonstrate the equivalence between clinical specimens and contrived specimens.

IVD, In vitro diagnostic; LoB, limit of blank, LoD, limit of detection, LoQ, limit of quantitation.
[a]The list is not exhaustive and is intended to provide a general overview and should not be interpreted as formal definitions.

to provide evidence of validity for the proposed indications for use. The ability to detect a biomarker accurately and reliably is dependent on the methodology; however, whether the results of the test are meaningful for a proposed indication is dependent on the target population and the cut-off or clinical decision point used to determine the clinical status (e.g., positive or negative for the disease). For example, the optimal cut-off used to evaluate the sensitivity and specificity of the test may be different in asymptomatic vs. symptomatic patients or may be acceptable for screening but not for diagnosis. For this reason, clinical validation study designs are specific to the clinical setting and proposed use enables the FDA to understand the benefits and risks of the test and determine if the test is safe and effective for the proposed use.

The main goals of a successful clinical validation study are to have a representative target population and to minimize sources of bias to overcome uncertainty in the veracity of the study conclusions. It is also important to understand how the test will fit into the current management of patients. This information informs whether the outcome of interest in the study is appropriate and whether the sensitivity, specificity, and predictive values[4] of the test are expected to improve patient management. A test that neither performs better than tests in current clinical practice nor contributes to refining the assessment of the patient status may lead to patient mismanagement.

Many test developers seek to clinically validate their test using previously collected, banked specimens. Such retrospective studies are feasible for most indications provided the test developer has an appropriate specimen selection protocol with prespecified inclusion and exclusion criteria for both the patient population and the specimen. Convenience sampling, which is the use of specimens based on being readily available, can be subject to significant bias and fail to represent the target population for various reasons. For example, samples may be missing from a banked set of tumor specimens because they were depleted due to prior repeat testing. The reasons for repeat testing may be those that would impact test performance such as small tumor volumes, failure to yield accurate results, or rare biomarkers, which were used in other studies. To address this type of bias, sponsors may design prospective-retrospective study designs by collecting specimens from all patients consecutively tested or enrolled between two time points, accounting for the missing samples, and demonstrating that variables that may impact test performance are balanced between the evaluable and unevaluable specimens [17]. These variables include patient demographics (e.g., age, gender, race/ethnicity, disease

---

[4] Sensitivity is estimated as the proportion of subjects with the target condition in whom the test is positive. Specificity of the test is estimated as the proportion of subjects without the target condition in whom the test is negative. Positive predictive value refers to the proportion of test-positive patients who do have the target condition, and negative predictive value refers to the proportion of test-negative patients who do not have the target condition [16].

stage, grade, comorbidities, and prior treatments outcomes) and specimen characteristics in some cases (e.g., tumor size, volume, and results). The variables of interest are largely determined by the indication. For example, test developers collecting specimens to support a prognostic indication can provide information about other prognostic variables to determine that the test is not biased to a unique patient population. Additional considerations may be given to often overlooked factors such as age of specimens and batch effects. For example, gene signature tests developed using specimens collected a decade earlier may not be representative of specimens obtained with patients treated with newly approved drugs and current guidelines. For these reasons, well-annotated specimen collection is important to provide evidence that the biomarker measurement is appropriately correlated with the outcome of interest but also that the sensitivity and specificity observed are representative of the test performance when commercialized in the real-world clinical setting.

Case-control studies are generally not considered appropriate for indications such as early detection and screening. This is because case controls are taken from patients who have already been diagnosed with cancer and are therefore not considered representative of the intended use population. Similarly, the use of healthy case controls may fail to include challenging cases such as patients with symptomatic but nonmalignant conditions. The use of case controls can lead to inflated estimates test performance for screening tests. Because positive cases may be rare, such as in screening of cancer, use of enrichment strategies provided the statistical analysis plan accounts for the enrichment and adjusts the analyses accordingly. A well-developed prespecified specimen selection plan for clinical studies may minimize sources of bias. Patient samples are adequately annotated with information that provides information the test performance across various disease stages and conditions.

A statistical analysis plan that describes the inclusion and exclusion criteria, study design, the primary and secondary endpoints, and method of calculating results. As noted earlier, the analysis plan may include, as applicable, demonstration of performance across disease stages and conditions, demonstration that subgroups can be pooled (such as when studies are conducted in both the US and non-US sites), demonstration that each biomarker in a multibiomarker test contributes to the effectiveness of the test, demonstration that the test performance is not biased to a select demographics, evaluation of the influence of pretest probabilities such as prior testing or treatments, and evaluation that the test results are not influenced by the underlying patient characteristics that are over- or underrepresented in the test set. For example, a test designed for the early detection of liver cancer may not be representative of the future test population if none of the patients in the study were positive for hepatitis B virus. Test validation may include the range of conditions that would be encountered in a clinical setting for which the use of the test would be applied.

When planning the statistical analysis plan, it is important to prespecify the performance needed to establish the test is clinically meaningful for the proposed use. This includes both the point estimate and the lower bound of the 95% confidence interval to determine the minimum number of subjects that would be enrolled into the trial, and accounting for those patients/specimens that may drop out of the study [17]. An appropriately sized study helps to minimize uncertainty in the observed performance. Additionally, the positive predictive value (PPV) and negative predictive value (NPV) are critical endpoints to assess, particularly for rare biomarkers where the low prevalence in the population can greatly influence the estimates of test performance. It is important to note that clinical sensitivity is the percentage of patients with a disease condition for whom the test will yield a positive result, and therefore, for example, a test with 100% sensitivity and 99.9% specificity may appear to be a very high-performing test. However, in a low prevalence setting, the significance of false positives is masked by the large number of negatives, and therefore, PPV is critical to understanding the potential for false-positive results. In other words, given the test result is positive, how likely is it that the patient has the condition of interest. A test with 100% sensitivity and 99.9% specificity may only have a PPV of 50% for a disease that has a prevalence of 1 in 1000 (see Table 3.3 as an example). A PPV of 50% means that for every person for whom cancer was correctly detected, the same number of patients were assigned false-positive results. In the cancer setting, these performance estimates are important to determine the benefit risks of the test (e.g., risk of procedures to follow up a false-positive result vs. the benefit of detecting cancer at a time where outcomes may be improved).

The performance estimates of sensitivity, specificity, PPV, and NPV are requested in support of validation. The FDA generally does not use receiver operative curves (ROCs) to validate IVD performance. Rather, a prespecified clinical threshold (i.e., cut-off, cut-point, and clinical decision point) is used to evaluate performance. For tests with intermediate results

Table 3.3 Example performance estimates in rare disease settings.

	Clinical truth Disease	No disease	
Positive test result	10	10	PPV = 50%
Negative test result	0	9990	NPV = 100%
	Sensitivity = 100%	Specificity = 99.9%	

or equivocal results (e.g., gray zones), prespecifying how the intermediate results will be counted (e.g., as negative or positive) depending on the indication and likely patient management in the proposed clinical setting for the purpose of conducting the analysis of performance. Sensitivity analyses are performed to estimate the "worst case scenario" for invalid test results [16].

Modifying the clinical thresholds based on the results of the clinical validation study is not appropriate; otherwise, the data set becomes a device "training" data set. Similarly, there are concerns about splitting data from a single study into training and validation data as the test may be biased to the study cohort. Independent validation ensures that the study is not biased. Under all conditions, data integrity should be maintained where testers are kept blinded to associated outcome data [16,17].

Review of personalized medicine diagnostics
Companion diagnostics

A companion diagnostic is an IVD device that provides information that is essential for the safe and effective use of a corresponding therapeutic product. The use of a companion diagnostic with a therapeutic product is stipulated in the instructions for use in the labeling of both the companion diagnostic and the corresponding therapeutic product [18]. Companion diagnostics are generally developed contemporaneously with the therapeutic indication, and recommendations for codevelopment have been outlined across multiple resources, including a draft guidance document titled Principles for Codevelopment of an In Vitro Diagnostic Device with a Therapeutic Product[5] [19] that describes details for planning for an effective companion diagnostic test strategy.

---

[5] See footnote 3.

The first FDA approved companion diagnostic was the Dako HercepTest, which is an immunohistochemistry test used to identify breast cancer patients with Her2 expression levels that are associated with response to HERCEPTIN (trastuzumab; Genentech), and now there are over 150 references to companion diagnostics/drug indications [20]. The rapid expansion of biomarker-based drugs has evolved to include group labeling claims whereby a single companion diagnostic can be labeled for a specific group of therapeutic products when certain conditions are met [21]. These conditions establish criteria by which the safety and efficacy of the drug are supported by the appropriate validated companion diagnostics. One example of a group of drugs for which a group indication was granted for select companion diagnostics is the tyrosine kinase inhibitors indicated for the treatment of patients with nonsmall-cell lung cancer (NSCLC) whose tumors have EGFR exon 19 deletions or exon 21 (L858R) substitution mutations. A group labeling indication enables the FDA-approved device to obtain indications for the entire group (e.g., example, afatinib, erlotinib, gefitinib, osimertinib, and dacomitinib).

Validation of companion diagnostics begins with pretrial planning to include a specimen banking and annotation strategy, including the careful consideration of stability of the specimen; for example, some immunohistochemistry biomarkers may degrade over time when stored as cut sections on slides. This is because most clinical trials, especially those with rare biomarkers, are accruing and enrolling patients based on results obtained with local laboratory testing or a central test (referred to as clinical trial assay or CTA) that is not intended as the final companion diagnostic (CDx), necessitating the need for a "bridging" study. Bridging studies are performed to support the safety and effectiveness of the companion diagnostic, which will be labeled for use with the corresponding therapeutic indication (i.e., demonstrate that the therapeutic product efficacy is maintained when the CDx is used to test the clinical trial patient samples). Bridging studies include both concordance to the clinical trial assay(s) using specimens enrolled in the pivotal trial and an evaluation of efficacy based on the test results obtained with the companion diagnostic. The bridging study should demonstrate that the companion diagnostic supports the safety and efficacy performance of the therapeutic product observed in the pivotal trial using the same endpoints as were used to evaluate the therapeutic indication [22,23].

Typically, clinical trials are targeted trials in which patients are only enrolled based on a biomarker-positive status. Although there are outcome data available for the CTA-positive enrolled patients, there are no outcome data for CTA-negative patients who were not enrolled. Therefore, it is important to obtain CTA negatives from the same target population to test with the companion diagnostic to obtain estimates of the CTA-negative and CDx-positive discordance and the efficacy results on negative subjects by conducting a sensitivity analysis based on the discordant results. The number of specimens is influenced in part by the prevalence of the biomarker. Bridging studies are challenged when there are multiple CTAs used in the trial due to differences in test methods, designs, limitations, or cut-offs. Additionally, when the biomarker is based on a specific definition, for example, any mutation in a select gene determined to be activating, the detection of the biomarker for local tests may be subject to different definitions and interpretations of activating mutations, all of which leads to potential discordance when specimens are tested with the companion diagnostic. For this reason, the pharmaceutical drug sponsor may be recommended to prespecify the definition of the biomarker used for enrollment [19,22].

Trials that stratify on the biomarker, such as different levels of protein expression as measured by an IHC assay, easily enable the reassessment of outcome based on the market-ready IHC device; however, it is not uncommon to have a large number of missing samples, particularly for those cancer types with very small amounts of tissue available (e.g., cholangiocarcinoma). When the percentage of missing samples is significant, there may be options for whether such data can be obtained in the post-approval setting using real-world evidence (RWE). For example, the FoundationOne CD assay (F1CDx, Foundation Medicine, Inc.) was approved with a postmarket condition of approval to obtain clinical outcome data such as RWE to confirm the clinical effectiveness of the F1CDx as a companion diagnostic device for the identification of patients with solid tumors with NTRK1/2/3 fusions and NSCLC patients with ROS1 fusions who may benefit from treatment with ROZLYTREK [24].

The FDA has approved tests that are IVDs that identify a subgroup(s) of the indicated patient population that has a different benefit/risk profile than the broader population for whom the corresponding therapeutic product is indicated (see Table 3.4). These tests don't meet the definition of a companion diagnostic but are nonetheless beneficial for clinical decision-making by predicting response based on the biomarker and information is included in the drug product labeling. In some cases, the

Table 3.4 Examples of in vitro diagnostics (IVDs) used to provide information about drug benefit based on a biomarker.

Diagnostic	Indication	Submission number[a]
Ventana PD-L1 SP142 IHC	PD-L1 expression in ≥ 50% TC or ≥ 10% IC as detected by VENTANA PD-L1 (SP142) Assay in NSCLC may be associated with enhanced overall survival from TECENTRIQ (atezolizumab).	P160006
Dako PD-L1 IHC 28–8 pharmDx	PD-L1 expression (≥1% or ≥5% or ≥10% tumor cell expression), as detected by PD-L1 IHC 28–8 pharmDx in nonsquamous NSCLC, may be associated with enhanced survival from OPDIVO.	P150025
	PD-L1 expression (≥1% tumor cell expression), as detected by PD-L1 IHC 28–8 pharmDx in UC, may be associated with enhanced response rate from OPDIVO.	P150025/ S003
	PD-L1 expression (≥1% tumor cell expression), as detected by PD-L1 IHC 28–8 pharmDx in SCCHN, may be associated with enhanced survival from OPDIVO	P150025/ S003

[a]Refers to the PMA submission number that can be searched in the PMA database https://www.accessdata.fda.gov/scripts/cdrh/cfdocs/cfpma/pma.cfm.

future data (e.g., phase 3 trial data used to support full approval of the drug) has resulted in modification of the assay to a companion diagnostic. An example is shown in Table 3.4, when the intended use of the Ventana PD-L1 SP142 IHC for TECENTRIQ[6], was later changed to a CDx claim.[7]

Tissue agnostic diagnostic tests in oncology

Tremendous technology developments have informed the pathogenic mechanisms at work in the emergence of cancer. Many cancers share these same underlying mechanisms that are the basis for the development of drugs whose mechanism of action is the same across multiple cancers

[6] See the Summary of Safety and Effectiveness for P160006 and P160002/S006 (CDx indication), respectively.

[7] Refer to the approved therapeutic product labeling for the most up-to-date indications.

that share the same biomarker. Tissue agnostic indications define the disease based on the biomarker rather than the histology.

In 2017, the FDA approved the first tissue agnostic indication, for pembrolizumab, an antibody against programed death receptor-1 (PD-1) for the treatment of unresectable or metastatic microsatellite instability-high (MSI-H) or mismatch repair deficient (dMMR) solid tumors [25]. MSI is the result of replication errors in tandem repeats due to defects in the MMR genes that are responsible for DNA-mismatch repair. At the time of the clinical trial, MSI-H and dMMR testing were widely performed in laboratories using qPCR methods designed to detect mismatch in select loci, or immunohistochemistry tests to detect the presence of proteins from four genes MLH1, MSH2, PMS2, and MSH6 in newly diagnosed patients with colorectal and endometrial cancer to identify patients who may be appropriate for further testing to determine if the patient has Lynch syndrome, which is associated with a predisposition to cancer. Although both MSI-H and dMMR tests were used for enrollment and there are a number of papers demonstrating reasonable concordance between the PCR- and IHC-based assays to support the identification of the target population after drug approval, differences in reporting due to unique mutations in the MMR genes can occur. A postmarketing commitment to support both PCR- and IHC-based tests for the proposed use was established. Next-generation sequencing (NGS) approaches were also developed due to the perceived advantage of this technology over other technologies, which included the ability to use less tissue, and comprehensive testing of hundreds of microsatellite loci. In partnership with Foundation Medicine, Inc. (Cambridge, MA), a bridging study was performed to support the F1CDx as a companion diagnostic. The reasonable assurance of safety and effectiveness for F1CDx for the detection of MSI-H status in patients with solid tumors who may benefit from treatment with pembrolizumab was established through a retrospective clinical device bridging study using tumor tissue FFPE specimens from patients enrolled in the Merck clinical studies KEYNOTE158 and KEYNOTE-164, and an additional set of commercially procured (tumor bank) specimens evaluated as a supplementary source of specimens for the biomarker. Additionally, to support the tissue agnostic indication, analytical performance was assessed in a variety of tumor types [26].

Another novel tissue agnostic biomarker is "tumor mutational burden." Tumor mutational burden (TMB) refers to the number of mutations found in the DNA of cancer cells. Cancers with high TMB are targets for immunotherapeutics because the mutations lead to novel antigens that have the potential to activate immune responses. The operational definition of TMB used to identify patients is mutations per megabase. The FDA approved KETYRUDA (pembrolizumab) on June 16, 2020, for the treatment of adult and pediatric patients with unresectable or metastatic TMB-high [TMB-H; ≥ 10 mutations/megabase (mut/Mb)] solid tumors that have progressed following prior treatment and who have no satisfactory alternative treatment options [27]. A companion diagnostic indication for high TMB at the cut-off of 10 mutations per megabase (mut/Mb) in patients with solid tumors who may benefit from treatment with pembrolizumab was also approved [28]. Notably, TMB scores are dependent on the test type (targeted panels and whole exome sequencing), and optimal scores for response may be cancer specific. For this reason, Friends of Cancer Research is one organization that is leading the TMB Harmonization Project to help reduce sources of variability across tests.

To date the FDA has approved a total of four oncology drugs for patient populations defined by the presence of a biomarker in tumor tissue rather than the cancer type (Table 3.5) [29]. The safety and efficacy of these tissue agnostic therapeutics are dependent on the availability of accurate companion diagnostic tests to test patient specimens. Analytical validation includes focus on a variety of tissue types that may present test challenges due to differences in specimen types or sources of interference such as melanin or necrotic tissue and sources of tumor heterogeneity. Clinical validation is supported by the data obtained from the drug trials.

Comprehensive genomic profiling tests

Medicine is being transformed by the availability of comprehensive genetic/genomic diagnostic tests because these tests can provide information about a multitude of possible actionable mutations with one test, expediting optimal decision-making and avoiding exhausting the patient specimen by having to run multiple tests. Advances in the analytical capabilities of these technologies, such as increased sensitivity and accuracy across challenging genomic contexts, have enabled diagnostic test developers to gain traction in broader types of specimens (e.g., plasma cfDNA and single cells) and biomarkers such as epigenetic modifications. These tests pose a challenge to the traditional paradigm for regulatory review and authorization, which is based on the analytical and clinical validation of each biomarker.

Table 3.5 Tissue agnostic drug approvals and companion diagnostics.

Drug	Biomarker	Companion diagnostic
Pembrolizumab (KEYTRUDA)	Microsatellite instability-high (MSI-H) or mismatch repair-deficient (dMMR) High tumor mutational burden (TMB $\geq$ 10 mutations per megabase)	FoundationOne CDx (F1CDx, Foundation Medicine, Inc.)
Dostarlimab (JEMPERLI)	Mismatch repair-deficient (dMMR)	VENTANA MMR RxDx Panel (Ventana Medical Systems/Roche Tissue Diagnostics)
Larotrectinib (VITRAKVI)	NTRK gene fusions NTRK1, NTRK2, and NTRK3	FoundationOne CDx (F1CDx, Foundation Medicine, Inc.)
Entrectinib (ROZLYTREK)	NTRK gene fusions NTRK1, NTRK2, and NTRK3	FoundationOne CDx (F1CDx, Foundation Medicine, Inc.)

MSI-H, Microsatellite instability-high; *NTRK*, neurotrophic receptor tyrosine kinase.

The authorization of the Memorial Sloan Kettering (MSK) IMPACT (Integrated Mutation Profiling of Actionable Cancer Targets) tumor profiling test in November 2017, a 468 gene panel, established a new Class II regulation (21 CFR 866.6080) for NGS tests used in oncology to identify somatic mutations in tumor tissues from patients with solid neoplasms [30]. The authorization enabled the development of special controls that are required to ensure quality NGS assays for proposed tumor profiling use. The unique analytical validation for these comprehensive panels was based on a representative approach to validation for variant types such as single-nucleotide variants (SNVs) and insertions and deletions (indels). While NGS tests used for companion diagnostics remain subject to the biomarker-specific analytical validation, a representative approach is one in which validation is provided based on the variant type or gene using representative specimens that include clinically relevant variants and variants in challenging regions. Validation is presented per variant, per variant type, and panel-wide where, for the latter, the mean of the data is reported across variant types and challenging genomic contexts such as high GC content, homopolymeric regions, or repetitive sequences. In addition, data that demonstrates the technical quality metrics such as base quality scores, coverage, minimum read depths, etc., can support high accuracy. Key quality metrics are then used as additional empirical evidence of the performance of the assay [31,32].

The authorization of the MSK-IMPACT used historical performance data obtained with clinical patient specimens in real-world use to support a pan-tumor indication. Test invalid rates at key steps in the procedure (e.g., library preparation and sequencing) were summarized for many different tumor types. In addition, data to support tumor profiling assays includes the processes used to curate, interpret, and establish the clinical significance of the variants as having either clinical significance or potential clinical significance. Variants are reported into distinct levels (tiers) based on the amount of evidence associated with the variant. Finally, NGS tests are dynamic tests in that they may undergo continuous improvements. FDA authorization of the MSK IMPACT included acceptance of prespecified testing and performance criteria used to make select improvements to the test. These modifications include information on the risks associated with the modification and whether the validation appropriately addresses that the high-quality performance is maintained.

The FDA has authorized several different types of comprehensive genetic tests for various intended uses that are outside the scope of the discussion presented here; however, a list of examples is provided in Table 3.6. The special controls for these tests can be found online in the classification order as well as the CFR database.[8] Finally, the FDA guidance document, "Considerations for Design, Development, and Analytical Validation of Next Generation Sequencing (NGS)-Based IVDs Intended to Aid in the Diagnosis of Suspected Germline Disease" provides recommendations for the validation of NGS-based tests used to diagnose patients.[9]

Liquid biopsies

Liquid biopsy refers to the blood-based detection methods for by-products of cancer such as circulating tumor cells and circulating tumor DNA (ctDNA) shed from tumors in patients. The advantages of liquid biopsies are wide in scope. For example, they offer a noninvasive method for providing information about whether a person's cancer has clinically

---

[8] Code of Federal Regulations Title 21 (CFR) available at https://www.accessdata.fda.gov/scripts/cdrh/cfdocs/cfcfr/cfrsearch.cfm and https://www.ecfr.gov/
[9] Available at https://www.fda.gov/media/99208/download

Table 3.6 Examples of in vitro diagnostics with comprehensive genetic results.

Device	Purpose	Regulation
Affymetrix CytoScan Dx Assay and Agilent GenetiSure Dx Postnatal Assay	Postnatal detection of CNVs associated with developmental delay, intellectual disability, congenital anomalies, or dysmorphic features	21 CFR 866.5920 Postnatal chromosomal copy number variation detection system
23andMe Personal Genome Service (PGS) Carrier Screening Test	Genetic test for autosomal recessive variants in adults	21 CFR 866.5940 Autosomal Recessive Carrier Screening Gene Mutation Detection System
Helix Laboratory Platform	Exome sequencing and detection of SNVs and small insertions and deletions (indels) in human genomic DNA extracted from saliva samples	21 CFR 866.6000 Whole exome sequencing constituent device

relevant somatic mutations, particularly when tumor tissue is not available, and they are becoming increasingly relevant in applications such as measurable residual disease and risk of cancer recurrence, potentially providing a window into the effectiveness of treatment, and detection or surveillance for cancer recurrence. Obtaining this information using traditional invasive surgical methods or imaging methods is limited to the tissue sampled and the resolution of the technology.

The first FDA-approved liquid biopsy tests were qualitative real-time PCR tests for specific companion diagnostic biomarkers. The cobas EGFR Mutation Test v2 (Roche Molecular Systems, Inc., Pleasanton, CA), a real-time PCR test for the detection of exon 19 deletions and L858R substitution mutations from cell-free, circulating tumor DNA (cf/ctDNA[10]) isolated from plasma, was approved as an aid in selecting patients for treatment with erlotinib (TARCEVA) [33]. This was quickly followed by FDA approval of the cobas EGFR Mutation Test v2 for the detection of EGFR exon 20 T790M substitution mutation as an aid in

---

[10] References to circulating free (cf) and circulating tumor (ct) DNA are made based on the terminology used by the sponsor but are referred to interchangeably.

the selection of patients for osimertinib (TAGRISSO) [34]. The clinical validation data with these tests highlighted two common clinical challenges with liquid biopsies. The first is that the ability to detect mutations is dramatically impacted by the shedding of ctDNA from the tumor into circulation. Poor shedding of ctDNA, either through apoptosis or necrosis, will lead to false-negative results due to the lack of detectable ctDNA. The instructions for use with the cobas test indicate that when considering erlotinib for patients with EGFR mutations based on plasma, patients who are negative by the cobas test should be reflexed to routine tissue biopsy and testing for EGFR mutations with the FFPE tumor tissue. The same language was used with the approval of the therascreen PIK3A RGQ PCR Kit for use in breast cancer patients considering the treatment with alpelisib [35].

The second challenge is that liquid biopsy tests may be detecting mutations in ctDNA shed from distant metastatic sites that are not confirmed when testing a single section from a primary tumor specimen due to the heterogeneous nature of tumor specimens and the development of new mutations in metastatic tissues. Because these tests are designed to have very high analytical sensitivity, and because biomarker-based clinical trials may not have outcome data in the biomarker-negative subset, the truth as to whether the liquid biopsy result is a false-positive cannot be determined in the absence of clinical outcome data in tissue-negative, ctDNA-positive patients. For this reason, recommendations are that plasma DNA testing is performed for patients for whom a tumor biopsy cannot be easily obtained.

In 2020, the first comprehensive NGS test to receive tumor profiling claims, as well as companion diagnostic claims for cfDNA from plasma, was the Guardant360 CDx (Guardant Health, Inc., Redwood CA) [36]. The test was approved for the detection of SNVs, indels, select copy number amplifications, and select fusions in a 55 gene panel. Clinical reporting of comprehensive mutations in ctDNA followed a tiered approach based on both the clinical significance of the biomarker and the level of analytical validation that was performed for the biomarker.[11] To support the safety and effectiveness of liquid biopsy tests for ctDNA, similar analytical validation as for tumor-tissue-based NGS tests were performed as described earlier; however, additional studies specific to ctDNA

---

[11] See Table 3.3 Category of Definitions in the Summary of Safety and Effectiveness (P200010) [36].

evaluated the unique preanalytical variables to which liquid biopsy tests are subject. This includes validation of the specific blood collection tubes used with the liquid biopsy assay because it has been shown that different blood collection tubes can lead to differences in the accuracy and reliability of cfDNA collection. Additional challenges with analytical validation of liquid biopsy assays include specimen stability, reproducibility, and the detection of clonal hematopoiesis of indeterminate potential (CHIP) variants. The broader liquid biopsy community such as the Blood Profiling Atlas in Cancer (BloodPac) consortium[12] and the International Liquid Biopsy Standardization Alliance (ILSA) Collaborative Community[13], are devoted to the development of analytical validation protocols to support the development, validation, and harmonization of liquid biopsy assays as they become instrumental for multiple uses such as molecular residual detection [37].

Reference materials and quality controls for genetic/genomic assays

NGS tests can enable an individual's genome to be surveyed for any one of thousands of variants associated with genetic disorders. It is difficult for test developers to obtain controls for these variants, and the absence of adequate reference materials makes it harder to assure accuracy of available tests. Thus, well-characterized reference materials and quality controls are fundamental to the laboratory and test developer community.

To support high-quality data for NGS assays, reference standards and controls are critical. Test developers include external controls for tests such as companion diagnostics to inform users when instrument or reagent integrity is compromised and prevent reporting incorrect results. For molecular diagnostics that detect mutations, it is important to use controls that represent various types of mutations (e.g., SNVs, indels, fusions, and CNVs), as well as mutations detected at a low percentage (low allele fraction) to be sensitive to deviations in test performance. There are several initiatives to support the availability of reference materials to help harmonize tests for critical biomarkers. The Medical Device Innovation Consortium (MDIC) Somatic Reference Sample initiative[14] and the Foundation for the National Institutes of Health (FNIH),[15] which

[12] https://www.bloodpac.org/

[13] https://fnih.org/our-programs/biomarkers-consortium/programs/ilsa

[14] https://mdic.org/project/cancer-genomic-somatic-reference-samples/

[15] https://fnih.org/our-programs/biomarkers-consortium/programs/ctdna-reference-materials

are multistakeholder public-private groups, aim to improve NGS testing by developing a set of reference samples for NGS-based oncologic tests with specific prioritization of actionable mutations. These quality control projects can help test developers standardize their NGS measurements so that accurate reporting is consistent across platforms and tests. The use of reference standards is also expected to support transparency among various test developers by informing analytical sensitivity of the test, which is critical to mitigating false-positive and false-negative results and making appropriate personalized medicine decisions.

Database recognition

In 2013, FDA cleared the Illumina MiSeqDx Cystic Fibrosis 139-Variant Assay and the Cystic Fibrosis Clinical Sequencing Assay, which were the first NGS assays to be authorized by the FDA [38,39]. Testing for variants in the cystic fibrosis transmembrane conductance regulator (CFTR) gene had expanded from 23 alleles recommended in 2004 by the American College of Medical Genetics (ACMG) to well over 100 variants or more depending on the test. These variants were based on real-world data (RWD) collected in the Clinical and Functional TRanslation of CFTR (CFTR2) database [40,41]. This database had established a curation process to ensure that pathogenic variants were supported by relevant CF-defining characteristics and laboratory test results that enabled the FDA to assess the clinical validity of the 139 variants and enable the expanded reporting across the gene as an aid in the diagnosis. The use of the CFTR2 database paved the way for enabling an adaptable approach to test the reporting of genetic variants.

NGS tests, like any diagnostic test, need valid scientific evidence to support the clinical validity of the findings. Sequencing technologies are generating exponentially large amounts of genetic information that require clinical interpretation of the specific variants identified in the nucleic acid isolated from a patient specimen. Such interpretations require evaluation of all levels of evidence such as, but not limited to, variant databases, literature including professional guidelines, patient experience found in medical records, laboratory tests, and in silico predictions for effects on protein function. Both public and proprietary databases have emerged to support this growing clinical need, as have awareness of the potential incongruities in the decisions regarding the variants in individual databases.

To accommodate the need to bridge clinical validation to test results, FDA created the Database Recognition Program in 2018. FDA's guidance "Use of Public Human Genetic Variant Databases to Support Clinical Validity for Genetic and Genomic-Based In Vitro Diagnostics; Guidance for Stakeholders and Food and Drug Administration Staff" [42] describes this program and FDA's intention to recognize high-quality, publicly accessible databases to support the clinical validity of comprehensive panel tests for test manufacturers. The intention is to help assure the quality of data and assertions for a dynamic database as valid sources of accurate human variant interpretation data become available. Database recognition indicates that the FDA believes the data and assertions contained in the database can be considered valid scientific evidence.[16]

The recognition process is one in which the FDA reviews all aspects of the database, including, but necessarily limited to, procedures implemented for training of personnel, variant curation, evaluation, assignment and re-evaluation processes, maintaining data integrity, version control, transparency, and cybersecurity. The FDA also reviews and verifies that assertions are consistent with the established processes/rules for interpretation laid out by the submitter. FDA has recognized two databases (See Table 3.7).

The first recognition was ClinGen, a research initiative funded by the National Institutes of Health (NIH) and managed by the National Human Genome Research Institute (NHGRI). The scope of the recognition included germline variants for hereditary diseases where there is a high likelihood that the disease or condition will materialize given a deleterious variant (i.e., high penetrance). The FDA reviewed oversight and governance procedures of the ClinGen Variant Curation Expert Panels (VCEPs), as well as a review of the assertions within the database. A complete description of the review [43] and links to the FDA-recognized information are available online.[17]

Memorial Sloan Kettering Cancer Center's Oncology Knowledge Base (OncoKB) [44] was the first somatic human variant database to receive FDA recognition [45]. OncoKB is a comprehensive database of

---

[16] US.Food and Drug Administration. FDA Recognition of Public Huam Genetic Variant Databases. https://www.fda.gov/medical-devices/precision-medicine/fda-recognition-public-human-genetic-variant-databases#: ∼ :text = FDA%20recognition%20of%20a% 20database,clinical%20validity%20of%20their%20tests.

[17] Clinical Genome Resources (ClinGen) available at https://clinicalgenome.org/about/fda-recognition/

Table 3.7 Food and Drug Administration database recognitions.

Database	Scope of recognition (if applicable)
Clinical Genome Resource (ClinGen)	Germline variants for hereditary disease where there is a high likelihood that the disease or condition will materialize given a deleterious variant (such as high penetrance)
MSK Oncology Knowledge Base (OncoKB)	Tumor mutations of Level 2 (evidence of clinical significance) and Level 3 (potential clinical significance) biomarkers

tumor alterations and clinical information with levels of evidence. Because databases that provide therapeutic recommendations are beyond the scope of the recognition, FDA recognized a portion of the database that MSK incorporated and uses the reporting format authorized for tumor mutation profiling assays (e.g., mutations with evidence of clinical significance and mutations with potential clinical significance[18]). Because the FDA reviews the database's policies and procedures for obtaining and maintaining variant assertions, after recognition, a database can continue to expand upon variant information within the database using the validated processes, provided that the database is maintained according to the specifications under which it was recognized. Test developers can then use FDA-recognized databases to support the clinical validity of their test by incorporating the recognized databases in a premarket submission. In addition to supporting premarket submissions, the database recognition program is intended to increase data sharing and transparency, the goal of which is to support and enhance improvements in patient care.

Unique initiatives

The 21st century Cures Act included provisions to accelerate the availability of innovations to patients. Three notable provisions are highlighted here.

Breakthrough Devices Program

The Breakthrough Devices Program is a voluntary program that is designed to support rapid access of beneficial devices to patients. The

---

[18] CDRH's Approach to Tumor Profiling Next Generation Sequencing Tests available at https://www.fda.gov/media/109050/download

program offers test developers the opportunity to interact with the FDA's staff to obtain timely feedback on protocols. Devices are eligible if they meet both statutory criteria described in the guidance document [46]. The first criterion requires a determination of whether the device provides for a more effective treatment or diagnosis of a life-threatening or irreversibly debilitating human disease or condition. The second criterion requires that one of the following factors be met: The test represents a breakthrough technology, no FDA-approved or cleared alternative is available, the device offers significant advantages over existing approved or cleared alternatives, or device availability is in the best interest of patients. Test developers are asked to provide information that supports a reasonable expectation of both clinical and technical success. For IVD tests, this may be clinical feasibility data and analytical validation such as precision to demonstrate the robustness for the novel measurand. Test developers can interact with FDA using the presubmission process to obtain feedback regarding whether their indications for use and novel test can support a breakthrough designation request.

Real-world data and real-world evidence

Patient data is a rich resource to mine clinical experience with approved drugs and diagnostics. RWD and RWE are gaining traction in several regulatory settings, including monitoring for safety signals and clinical decision-making. RWD is the data relating to patient health status and/or the delivery of health care routinely collected from a variety of sources. RWE is the clinical evidence about the usage and potential benefits or risks of a medical product derived from the analysis of RWD. The FDA has experience using RWD in regulatory decision-making for diagnostic devices. This includes the CFTR2 database at the Johns Hopkins patient registry as a source of clinical validity for cystic fibrosis tests [40], internally acquired data such as a laboratory's internal experience with testing tumor tissues to inform test performance, and finally other databases, literature, and guidelines. The Oncology Center of Excellence's (OCE's) Oncology RWE Program, established in 2020, advances the appropriate use of RWE in oncology product development to facilitate patient-centered regulatory decision-making. Multiple stakeholders are involved in developing systematic and rigorous approaches to advance oncology products. Additionally, FDA published the 2017 guidance on the Use of RWE to

Support Regulatory Decision-Making for Medical Devices [47]. Successful examples are available online.[19]

Predetermined change control plans

Medical products incorporating artificial intelligence and machine learning necessitate options for sponsors to support the modifications while continuing to provide a reasonable assurance of device safety and effectiveness. Predetermined change control plans (PCCPs) hallmark a mechanism whereby protocols provided in marketing submissions can be reviewed and authorized for both Class II and Class III devices. PCCPs describe planned changes that may be made to the device (and that would otherwise require a supplemental application under Section 515 of the Federal Food Drug and Cosmetic Act) if the device remains safe and effective. A PCCP includes a description of the modification and the risks of those modifications, and what protocols will be followed when developing, validating, and implementing the modifications to assure the device remains safe and effective. Test developers considering PCCPs should interact with FDA to determine if this novel approach is an appropriate regulatory strategy for their product.

Interacting with FDA

The mission of the FDA is to protect and promote public health, and CDRH is committed to assuring that patients and providers have timely and continued access to safe, effective, and high-quality medical devices. To support this mission, CDRH provides comprehensive regulatory assistance. This includes the Division of Industry and Consumer Education (DICE),[20] online device advice that includes access to educational webinars and courses, databases, guidance documents[21], and outreach to stakeholders through various workshops, industry working groups, and town hall meetings. In addition, the voluntary "Q-Submission" program provides avenues for test developers to request feedback and meetings with FDA regarding potential or planned medical device applications and submissions. This includes the presubmission program that enables early

---

[19] See https://www.fda.gov/media/146258/download

[20] Division of Industry and Consumer Education (DICE) available at https://www.fda.gov/medical-devices/device-advice-comprehensive-regulatory-assistance/contact-us-division-industry-and-consumer-education-dice

[21] Device Advice: Comprehensive Regulatory Assistance available at https://www.fda.gov/medical-devices/device-advice-comprehensive-regulatory-assistance

interactions between the FDA and test developers to obtain FDA feedback on any variety of topics such as those discussed here. "Informational meeting" Q-submissions are also useful to introduce FDA review teams to novel devices and new technologies in advance of seeking presubmission feedback. The types of-Q submissions and processes are described in the guidance document "Requests for Feedback and Meetings for Medical Device Submissions: The Q-Submission Program" [48].

Final considerations

Cutting-edge technologies are providing greater information about the unique etiology, risk, prognosis, and treatment of disease, creating novel diagnostics that are transforming medical care and expanding the ability of the clinical community and industry to develop effective treatments that will work best for each patient. The challenge to the medical community is parsing this information into clinically actionable vs. research-level data as the data is becoming integrated into routine care. This chapter discussed principally qualitative, DNA-based genetic tests; however, diagnostic tests that evaluate RNA expression and patterns of gene signature have also been authorized by the FDA and to date have been most frequently used to stratify diagnostic subsets and provide prognostic information. These tests along with multiomic areas (e.g., proteomic, transcriptomics, and microbiome) of medical research are gaining traction and expanding our understanding of disease mechanisms but are challenging to validate as information becomes more granular and patient-centric (i.e., n of one). The ability to evaluate response to a therapeutic for each biomarker is not feasible, and the need to accelerate availability of potentially helpful therapeutics is a priority. To promote health care, the FDA has developed regulatory strategies to advance patient access to new technologies and provide evidence of safety and effectiveness. These included representative validation approaches, tumor-agnostic indications, database recognition, and RWD use. In conclusion, high-quality patient health care starts with high-quality diagnostics. The FDA works to ensure that innovative advances in diagnostics are expedited to patients and their physicians so that patients and clinicians can receive accurate and clinically meaningful test results in support of successful patient outcomes.

Acknowledgment

The author wishes to thank Zivana Tezak and Pamela Ebrahimi for their review.

References

[1] U.S. Food & Drug Administration. Precision medicine, September 27, 2018. https://www.fda.gov/medical-devices/in-vitro-diagnostics/precision-medicine

[2] Subbiah V, Wirth LJ, Kurzrock R, Pazdur R, Beaver JA, Singh H, et al. Accelerated approvals hit the target in precision oncology. Nat Med 2022;28(10):1976−9.

[3] U.S. Food & Drug Administration Code of Federal Regulations Title 21§809.3(a). Definitions.

[4] U.S. Food & Drug Administration Code of Federal Regulations Title 21§862. 1300 Follicle-Stimulating Hormone Test System.

[5] U.S. Food & Drug Administration Code of Federal Regulations Title 21§866.5940. Autosomal recessive carrier screening gene mutation detection system.

[6] U.S. Food & Drug Administration. De Novo Classification Process (Evaluation of Automatic Class III Designation) Guidance for Industry and Food and Drug Administration Staff, October 5, 2021. https://www.fda.gov/media/72674/download

[7] U.S. Food & Drug Administration. Evaluation of Automatic Class III Designation (De Novo) Summaries. https://www.fda.gov/about-fda/cdrh-transparency/evaluation-automatic-class-iii-designation-de-novo-summaries

[8] U.S. Food & Drug Administration. Humanitarian Device Exemption Resources. https://www.fda.gov/medical-devices/humanitarian-device-exemption/resources

[9] U.S. Food & Drug Administration. Part 820 Quality System Regulation. https://www.accessdata.fda.gov/scripts/cdrh/cfdocs/cfcfr/CFRSearch.cfm?CFRPart = 820

[10] U.S. Food & Drug Administration. Quality System (QS) Regulation/Medical Device Good Manufacturing Practices. https://www.fda.gov/medical-devices/postmarket-requirements-devices/quality-system-qs-regulationmedical-device-good-manufacturing-practices

[11] U.S. Food & Drug Administration. Investigational Device Exemptions. https://www.accessdata.fda.gov/scripts/cdrh/cfdocs/cfcfr/CFRsearch.cfm?CFRPart = 812

[12] U.S. Food & Drug Administration. Investigational IVD Used in Clinical Investigations of Therapeutic Products, Draft Guidance for Industry, Food and Drug Administration Staff, Sponsors, and Institutional Review Boards, December 18, 2017. https://www.fda.gov/media/109464/download

[13] U.S. Food & Drug Administration. Search for FDA Guidance Documents. https://www.fda.gov/regulatory-information/search-fda-guidance-documents

[14] U.S. Food & Drug Administration. Recognized Consensus. https://www.accessdata.fda.gov/scripts/cdrh/cfdocs/cfstandards/search.cfm

[15] U.S. Food & Drug Administration. Medical Device Databases. https://www.fda.gov/medical-devices/device-advice-comprehensive-regulatory-assistance/medical-device-databases

[16] U.S. Food & Drug Administration. Statistical Guidance on Reporting Results from Studies Evaluating Diagnostic Tests; Guidance for industry and FDA Staff, March 13, 2007. https://www.fda.gov/media/71147/download

[17] U.S. Food & Drug Administration. Design Considerations for Pivotal Clinical Investigations for Medical Devices Guidance for Industry, Clinical Investigators, Institutional Review Boards and Food and Drug Administration Staff. November 7, 2013.

[18] U.S. Food & Drug Administration. In Vitro Companion Diagnostic Devices Guidance for Industry and Food and Drug Administration Staff, August 2014 https://www.fda.gov/media/81309/download

[19] U.S. Food & Drug Administration. Principles for Codevelopment of an In Vitro Diagnostic Device with a Therapeutic Product. Draft Guidance for Industry and Food and Drug Administration Staff, July 14, 2016 https://www.fda.gov/media/99030/download

[20] U.S. Food & Drug Administration. List of Cleared or Approved Companion Diagnostic Devices (In Vitro and Imaging Tools), January 2023. https://www.fda.gov/medical-devices/in-vitro-diagnostics/list-cleared-or-approved-companion-diagnostic-devices-in-vitro-and-imaging-tools

[21] U.S. Food & Drug Administration. Developing and Labeling In Vitro Companion Diagnostic Devices for a Specific Group of Oncology Therapeutic Products Guidance for Industry, April 2020 https://www.fda.gov/media/120340/download

[22] Meijuan Li. Statistical consideration and challenges in bridging study of personalized medicine. J Biopharm Stat 2015;25(3):397—407.

[23] Denne JS, Pennello G, Zhao L, Chang S, Althouse S. Identifying a Subpopulation for a Tailored Therapy: bridging clinical efficacy from a laboratory-developed assay to a validated in vitro diagnostic test kit used to identify a subpopulation for a tailored therapy. Stat Biopharm Res 2014;6(1):78—88.

[24] U.S. Food & Drug Administration. Summary of Safety and Effectiveness P170019/S014, June 7, 2022. https://www.accessdata.fda.gov/cdrh_docs/pdf17/P170019S014B.pdf

[25] Lemery S, Keegan P, Pazdur R. First FDA approval agnostic of cancer site—when a biomarker defines the indication. N Engl J Med 2017;377:1409—12.

[26] U.S. Food & Drug Administration. F1CDx Summary of Safety and Effectiveness P170019-S029 https://www.accessdata.fda.gov/cdrh_docs/pdf17/P170019S029B.pdf

[27] Marcus L, Fashoyin-Aje LA, Donoghue M, Yuan M, Rodriguez L, Gallagher PS, et al. FDA approval summary: pembrolizumab for the treatment of tumor mutational burden-high solid tumors. Clin Cancer Res 2021;27(17):4685—9.

[28] U.S. Food & Drug Administration. F1CDx Summary of Safety and Effectiveness P170019-S029 https://www.accessdata.fda.gov/cdrh_docs/pdf17/P170019S016B.pdf

[29] Lemery S, Fashoyin-Aje L, Marcus L, Casak S, Schneider J, Theoret M, et al. Development of tissue-agnostic treatments for patients with cancer. Annu Rev Cancer Biol 2022;6(1):147—65.

[30] U.S. Food & Drug Administration. MSK-IMPACT Decision Summary https://www.accessdata.fda.gov/cdrh_docs/reviews/den170058.pdf

[31] Jennings LJ, Arcila ME, Corless C, Kamel-Reid S, Lubin IM, Pfeifer J, et al. Guidelines for validation of next-generation sequencing-based oncology panels: a joint consensus recommendation of the Association for Molecular Pathology and College of American Pathologists. J Mol Diagn 2017;19(3):341—65.

[32] New York Department of Health Next Generation Sequencing (NGS) Guidelines for Somatic Genetic Variant Detection, April 2021. https://www.wadsworth.org/sites/default/files/WebDoc/NextGenSeqONCOGuidelines%20_April_2021.pdf

[33] U.S. Food & Drug Administration. Summary of Safety and Effectiveness Roche Molecular Systems, Inc. cobas EGFR Mutation Test v2. (P150047). https://www.accessdata.fda.gov/cdrh_docs/pdf15/p150047b.pdf

[34] U.S. Food & Drug Administration. Summary of Safety and Effectiveness Roche Molecular Systems, Inc. cobas EGFR Mutation Test v2. (P150044). https://www.accessdata.fda.gov/cdrh_docs/pdf15/P150044B.pdf

[35] U.S. Food & Drug Administration. Summary of Safety and Effectiveness QIAGEN therascreen PIK3CA RGQ PCR Kit (P190004). https://www.accessdata.fda.gov/cdrh_docs/pdf19/P190004B.pdf

[36] U.S. Food & Drug Administration. Summary of Safety and Effectiveness Guardant Health Inc., Guardant360 Dx (P200010) https://www.accessdata.fda.gov/cdrh_docs/pdf20/P200010B.pdf

[37] Vellanki PJ, Ghosh S, Pathak A, Fusco MJ, Bloomquist EW, Tang S, et al. Regulatory implications of ctDNA in immuno-oncology for solid tumors. J Immunother Cancer 2023;11(2).

[38] U.S. Food & Drug Administration. The Illumina MiSeqDx Cystic Fibrosis 139-Variant Assay Decision Summary (k124006). https://www.accessdata.fda.gov/cdrh_docs/reviews/k124006.pdf

[39] U.S. Food & Drug Administration. Illumina MiSeqDxTM Cystic Fibrosis Clinical Sequencing Assay (k132750). https://www.accessdata.fda.gov/cdrh_docs/reviews/K132750.pdf

[40] The Clinical and Functional TRanslation of CFTR (CFTR2); available at http://cftr2.org.

[41] Sosnay PR, Siklosi KR, Van Goor F, Kaniecki K, Yu H, Sharma N, et al. Defining the disease liability of variants in the cystic fibrosis transmembrane conductance regulator gene. Nat Genet 2013;45(10):1160—7.

[42] U.S. Food & Drug Administration. Use of Public Human Genetic Variant Databases to Support Clinical Validity for Genetic and Genomic-Based In Vitro Diagnostics Guidance for Stakeholders and Food and Drug Administration Staff, April 13, 2018. https://www.fda.gov/media/99200/download

[43] U.S. Food & Drug Administration. Genetic Database Recognition Decision Summary for ClinGen Expert Curated Human Variant Data (Q181150), December 4, 2018.

[44] Chakravarty D, Gao J, Phillips SM, Kundra R, Zhang H, Wang J, et al. OncoKB: a precision oncology knowledge base. JCO Precis Oncol 2017; Jul;2017.

[45] U.S. Food & Drug Administration. Genetic Database Recognition Decision Summary for OncoKB; Q191007 https://www.fda.gov/media/152847/download

[46] U.S. Food & Drug Administration. Breakthrough Devices Program Guidance for Industry and Food and Drug Administration Staff, December 18, 2018. https://www.fda.gov/media/108135/download

[47] U.S. Food & Drug Administration. Guidance for Industry and Food and Drug Administration Staff; Examples of Real-World Evidence (RWE) Used in Medical Device Regulatory Decisions, August 31, 2017. https://www.fda.gov/media/99447/download

[48] U.S. Food & Drug Administaration. Requests for Feedback and Meetings for Medical Device Submissions: The Q-Submission Program, June 2023. https://www.fda.gov/regulatory-information/search-fda-guidance-documents/requests-feedback-and-meetings-medical-device-submissions-q-submission-program.

CHAPTER 4

Precision medicine: success stories and challenges from science to implementation

Attila A. Seyhan[1,2,3,4] and Claudio Carini[5,6]
[1]Laboratory of Translational Oncology and Experimental Cancer Therapeutics, Warren Alpert Medical School, Brown University, Providence, RI, United States
[2]Department of Pathology and Laboratory Medicine, Warren Alpert Medical School, Brown University, Providence, RI, United States
[3]Joint Program in Cancer Biology, Lifespan Health System and Brown University, Providence, RI, United States
[4]Legorreta Cancer Center, Brown University, Providence, RI, United States
[5]School of Cancer & Pharmaceutical Sciences, Faculty of Life Sciences & Medicine, New Hunt's House, King's College London, Guy's Campus, London, United Kingdom
[6]Biomarkers Consortium, Foundation of the National Institute of Health, Bethesda, MD, United States

Introduction

Precision medicine is a data-driven approach that integrates phenotypic, genomic, epigenetic, and environmental factors (such as diets) that are unique to an individual or a group of individuals. Precision medicine is primarily based on understanding the molecular sequence of critical biologic events of the disease [1]. The precision medicine strategy facilitates disease diagnosis, predicts effective therapies, and avoids adverse reactions like toxicity. Precision medicine and personalized medicine are terminologies often used interchangeably [2]. Precision medicine has been extensively used in oncology; it has also been used in rheumatology, pulmonology, and several other therapeutic areas. Although significant efforts have been achieved by implementing precision medicine to reduce the failures of clinical trials by analyzing possible links to the physicochemical properties of drug candidates, unfortunately, so far the results have been unconvincing due to the limited size of data sets [3].

Several pieces of evidence have been showing that the preclinical results are often misleading and irreproducible [4–6]. In addition, findings from recent clinical trials indicate that not every patient benefits from a precision medicine strategy. Recent findings demonstrate that precision medicine-driven trials are not effective in 93% of patients. This is mostly

The New Era of Precision Medicine
DOI: https://doi.org/10.1016/B978-0-443-13963-5.00008-X

© 2024 Elsevier Inc.
All rights reserved.

83

due to the disease pathogenesis and patient heterogeneity [7] as well as the quality and timing of the patient specimen collection, the pathological stage of the disease, and the lack of predictive biomarkers allowing patient stratification to targeted treatments contributing to this failure collectively.

The scientific community has now widely accepted the lack of reproducibility and translatability from animal data into human studies. However, the real conundrum of drug development lies in the high attrition rates and the lack of translatability of data from preclinical to clinical studies. The poor translatability resulted in only 6% of animal data being translatable into human studies [8]. Moreover, the lack of translatability is anticipated to grow as new targeted treatments are approved for a relatively small group of patients [9].

The real problem lays on the lack of reliable, robust, and validated translatable biomarkers to improve therapeutic success. To date, however, several types of biomarkers have been playing a significant role in drug development. The process of identifying and validating biomarkers remains a major challenge. Recent advances in multiple "omics" combined with bioinformatics, biostatistics, machine learning algorithms, and artificial intelligence (AI) have helped to accelerate the discovery and development of biomarkers. Biomarkers play a significant role in optimizing the decision-making process during drug development. The ability to translate preclinical into clinical biomarkers will pave the way toward the implementation of precision medicine across different therapeutic areas. However, due to the rarity of many therapeutic targets, focused testing to investigate each clinically relevant biomarker individually is unlikely to be cost-effective. Thus, comprehensive genomic profiling by multigene sequencing panels may be needed. Various national and international guidelines have now incorporated specific recommendations for the use of multigene panel testing in specific settings [9,10].

Why we need precision medicine now?

Precision medicine is an emerging approach that looks at the root cause of an illness rather than addressing symptoms alone. Digital platforms can take into account individual variability in genes, environment, and lifestyle for each person. The need for such platforms is driven by the unique genetic characteristics and heterogeneity of the population. Precision medicine offers significant opportunities to shape the future of health care in its ability to guide health care (Fig. 4.1).

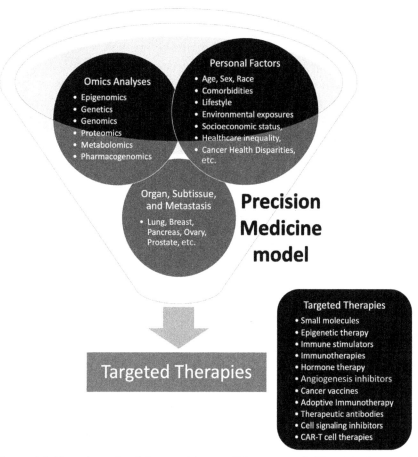

Figure 4.1 The schematic of the precision medicine approach. The precision medicine approach is characterized by treatments customized to specific gene mutations or genomic or other molecular alterations and personal factors of the patient relevant to each unique case of cancer. Companion diagnostics help guide in identifying targeted therapies that may be most effective for a specific patient or a group of patient's tumors [11]. *Adapted from Seyhan AA. The current state of precision medicine and targeted-cancer therapies: where are we? In: Scotti MT, Bellera CL, editors. Drug target selection and validation computer-aided drug discovery and design, vol. 1; 2022. p. 119–200; Krzyszczyk P, Acevedo A, Davidoff EJ, Timmins LM, Marrero-Berrios I, Patel M, et al. The growing role of precision and personalized medicine for cancer treatment. Technol (Singap World Sci) 2018;6:79–100.*

Precision medicine has been used for disease prevention, diagnosis, and treatment. Genetic mutations are the major contributors to cancer development and other diseases. The precision medicine approach has the potential to offer a better response rate to the developed therapeutic interventions taking into account the genetic/genomic alterations; hence, the success of therapy sparing patients from unnecessary toxicity. Precision medicine research is an interconnected system (Fig. 4.2). Clinical data and molecular profiling work together to generate a precise approach for the prevention, diagnosis, and treatment of cancer and other diseases.

Precision medicine relies on information to target specific genetic/genomic alterations responsible for the growth and spread of the tumor [12]. Those targeted therapeutics are designed to target cancer cells with

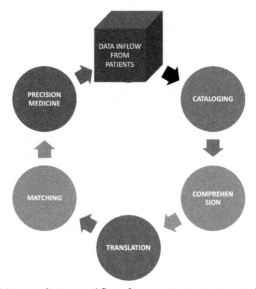

Figure 4.2 Precision medicine workflow from patients to targeted therapies: Each specific research discipline such as molecular profiling of a patient's or a group of patients' tumor genetic/genomic molecular profiling data, as well as individual's background, lifestyle, and physical characteristics, acts as the input, and this inflow of data from different sources sets the interconnected system into motion, thus contributing to the forward movement of precision medicine. It is of note that results from matching molecular alterations to targeted therapies can feed into the system as data inflow both for validating the approach and feeding new data. Each component can be categorized in individual entities such as *cataloging* the genetic alterations found in all cancer types, including genetic mutations, chromosomal rearrangements, and epigenomic changes; with a specific focus on identifying genetic differences between cancer subtypes; *comprehending* how the genetic characteristics of cancer contribute to cancer biology, and determining which characteristics are the most influential and clinically actionable; *translating* the findings related to genetic alterations and biological abnormalities of cancer into biomarkers, targeted therapies, and better treatment strategies; and *matching* each patient with the most effective and appropriate targeted therapies specific to a genetic/genomic molecular alteration. *Adapted from: https://ocg.cancer.gov/e-newsletter-section/precision-oncology.*

specific genetic/genomic aberrations, thus causing fewer side effects than chemotherapies [13]. In 2019, 39 out of 48 new drug approvals were granted to targeted therapies, and almost 25% of those approvals were cancer therapies [14]. At the same time, seven diagnostic tests were also approved in 2019 to stratify patients with specific genetic/genomic aberrations ("2019 Device Approvals." United States Food and Drug Administration, FDA.gov/medical-devices/recently-approved-devices/2019-device-approvals).

Several targeted therapies are now available, offering patients affected by different types of cancer better outcomes [15]. Several different drugs like imatinib targeting BCR-ABL showed remarkable efficacy in patients with chronic myeloid leukemia (CML) [16]. Another example is NTRK gene fusions that are present in multiple tumor types allowing NTRK-targeted therapies to be used for patients irrespective of the type of tumor histology [17]. The recent advances in molecular profiling technologies and computational tools have been unraveling molecular signatures that can be matched to specific target therapy [18]. With such a development, it is expected that the number of patients, who will benefit from the precision medicine approach, will increase significantly. Despite all those recent successes, the biggest challenge for precision medicine still remains howto interpret molecular profiling data and match the most appropriate targeted therapies for patients with those genetic/genomic alterations.

The paradox of precision medicine

Precision medicine sounds like an inarguably good thing to do. It begins with the observation that individuals differ in their genetic makeup and their response to treatments. Precision medicine besides genomics and other omics also incorporates a multiplicity of disciplines, including computational and bioinformatics. More recently, machine learning and AI have been gaining wider acceptance [19]. Precision medicine began in 1998 with the discovery of the *BCR-ABL* rearrangement in CML, which was successfully targeted by the drug imatinib. This resulted in a significant clinical remission, leading its approval in 2001 by the FDA [20].

Despite numerous successes, the notion of precision medicine has gained supporters and detractors within the scientific community. The supporters will quote ivacaftor, a drug that has eased symptoms in a small subset of patients with cystic fibrosis (CF). In contrast, detractors underlie ivacaftor, as a drug that took decades to develop, and is useless in 95% of patients whose mutations are different from the ones that ivacaftor acts on.

The same paradox applies to nearly every example of precision medicine we could find. However, clinicians viewed the use of a patient's genotype to determine the right dose of warfarin as a godsend until it was proved that this approach did not work any better than dosing through old-fashioned clinical measures (age, weight, and gender). Imatinib was hailed as an emblem of targeted cancer therapy when it shrank tumors in a subset of leukemia patients with a very specific mutation in their tumors. But then, in a lot of patients, tumors developed new mutations that made them resistant to the drug, and when they did, cancer returned. Although scientists have identified tens of thousands of genetic variations that could play a role in cancer and other diseases, ultimately even drugs that are a good match for a mutation don't always work. A targeted therapy that works in melanoma does not work in colorectal cancer even when patients have the same mutation.

The imprecision of precision medicine

By definition, precision medicine refers to the tailoring of disease prevention and treatment based on the characteristics of each patient. Until today, the debate on precision medicine in oncology and other therapeutic areas has revolved around whether it improves outcomes in our patients. Based on the available data, the answer is that it does work in a small group of patients. However, precision medicine has already achieved numerous successes in several diseases. Though on the one hand, it is not time yet for the precision medicine celebration. On the other hand, precision medicine deserves the credit for success since there are several positive examples in oncology and other diseases where success was achieved due to precision medicine.

Precision medicine has gained enormous popularity to improve the benefit/risk profile of the specific treatment. In reality, we don't seem to have achieved clear and tangible results from this approach in a large group of patients. Experts have questioned on several occasions the value of precision medicine. It may well be that the concept of precision medicine is rather imprecise. Instead, it would be more prudent to first look at biomarkers, because if there are errors in the biomarker measurement or questions about their validity, then the whole framework of precision medicine crumbles. Precision medicine is impossible without precision reliability in the measurement and validity of biomarkers. However, accurate biomarker (s) measurement is not easy with several potential avenues for errors.

The case history of excision repair crosscomplement group 1 protein (ERCC1) as a biomarker for platinum therapy in nonsmall-cell lung cancer (NSCLC) illustrates these problems very well. ERCC1 has been attempted to sell as a predictive biomarker for platinum therapy in NSCLC for many years and is now even available commercially, but no evidence supports or refutes this claim. The problem of validity in measuring biomarkers is relevant to modern immune checkpoint inhibitors, for example, PD-1 inhibitors such as nivolumab and pembrolizumab. Those drugs are truly useful in some cancers, but only in a subset of patients. For this subset, PD-1 inhibitors can sometimes provide durable responses, but for most patients, they do not provide much benefit. In fact, those drugs cause several side effects, including toxicity. Thus, we need biomarkers that can help to determine for this therapy which patients would benefit and which only experience toxicity. In the absence of predictive biomarkers, we are potentially exposing all eligible patients to toxicity.

In conclusion, there is a pressing need to reach a consensus on the measurement of biomarkers. Precision medicine to be effective requires precise measurement and validation of biomarker(s). Imprecise biomarker(s) measurement can lead to the erroneous conclusion that precision medicine is ineffective, while the issue could simply be the imprecision in biomarker measurement.

According to DeciBio's analysis of survey data and comparisons with 2018 data on the adoption and use of emerging precision medicine biomarkers and technologies in routine clinical care, five key themes emerged (https://www.decibio.com/insights/update-on-the-adoption-and-utilization-of-emerging-precision-medicine-biomarkers-and-technologies-in-routine-clinical-care).

1. Significant increase in the adoption of biomarker testing
2. Use of next-generation sequencing (NGS) as the standard method for molecular testing across cancer types; (however, other emerging biomarkers and modalities are not expected in the foreseeable future).
3. Adoption of liquid biopsies (LBx) and LBx is expected to be the most impactful change in pathology/biomarker testing.
4. Broader adoption of digital pathology use. Expectations for the impact of digital pathology remain high with the increasing development and use of AI.
5. And notably, the role of the pathologist is changing, and pathologists of the future will need to adjust to the changing landscape (and acquire a new skill set). It is noteworthy that in the last few years, there have

03-16-2018	CMS – LCD for NGS testing in advanced cancer patients
03-30-2018	Foundation Medicine – FoundationOne CDx (the 1st FDA approved CPG assay)
07-12-2018	Guardant – Medicare coverage of Gurdanat360 assay in NSCLC under Palmetto GBA, coverage later expanded to all solid tumors
01-07-2019	Guardant – LUNAR-1 MRD assay to biopharma and academic researchers
03-07-2019	Paige – Paige AI receives FDA breakthrough device designation for its AI-based digital pathology tools
12-19 – 01-19	ARCHER – ArhcerDx personalized MRD assay, receives breakthrough designation
04-01-2020	SECTRA, Leica – Sectra's and Leica's digital pathology solution for primary diagnosis approved by the FDA
04-21-2020	FDA - FDA expanded authorization for use of digital pathology during the COVID-19 pandemic
05-2020	Lilly and Novartis – Lilly's Retevmo and Novartis's Tabrecta approved for RET fusion and METex14 skipping in NSCLC, respectively
06-16-2020	MERCK, FOUNDATION MEDICINE – Keytruda approved by the FDA in advanced patients with TMB-H tumors with F1CDx as a CDx
07-30-2020	AstraZeneca – Tagrisso granted breakthrough designation for adjuvant treatment of patients with stage IB-IIIA EGFR-mut NSCLC
07-2020	GUARDANT HEALTH, FOUNDATION MEDICINE – Guardant's Guradant360, then Foundation Medicine's F1Liquid CDx become the first FDA-approved liquid biopsy CDx tests
09-2020	CMS, NATERA - CMS grants LCD for Natera's Signatera for CRC MRD and a draft LCD for immunotherapy monitoring
10-13-2020	CARIS LIFESCIENCES – Caris launches its CODEai real-world data clinicogenomic database with >215,000 patients
2021	• NCCN Clinical Practice Guidelines in Oncology - recommendations for sequencing of ALK, ROS1, EGFR, BRAF, KRAS, MET, RET, and NTRK for various solid tumors
03-2021	FDA approval of Lorlatinib for ALK positive NSCLC as first line therapy

Figure 4.3 Select key precision medicine developments; 2018−21 YTD. *Adapted from https://www.decibio.com/insights/update-on-the-adoption-and-utilization-of-emerging-precision-medicine-biomarkers-and-technologies-in-routine-clinical-care and; Mateo J, Steuten L, Aftimos P, Andre F, Davies M, Garralda E, et al. Delivering precision oncology to patients with cancer. Nat Med 2022;28:658−65.*

been several clinical, regulatory, technological, and commercial developments that have impacted precision oncology (Fig. 4.3).

What are the challenges and solutions with precision medicine?

Precision medicine is a growing field. Many of the technologies that are needed to meet the goals of precision medicine have only recently been developed. Cost is also an issue with precision medicine. Additionally, drugs that are developed to treat conditions based on molecular or genetic variations are likely to be expensive. Reimbursement from third-party payers for these targeted drugs is also likely to become an issue.

If precision medicine approaches are to become part of routine health care, doctors and other healthcare providers will need to know more about molecular genetics, biochemistry, and immunology. Despite recent breakthroughs and the growing force behind precision medicine, there are still substantial challenges and barriers to overcoming the broad precision medicine implementation into medical practice.

The precision medicine challenges to overcome include the following:

1. More research is needed to make a case for precision medicine adoption, which would lead to significantly improved outcomes.
2. Significant resources are needed to improve data collection, storage, sharing, and integration with EHR (electronic health records).
3. Incorporating genomic information into clinical care and research is not yet achieved and will be crucial in many healthcare systems.
4. Education is a key challenge because many physicians express little confidence in their ability to make clinical decisions when genetic or genomic information is available.
5. Current deterrents for complete patient participation are patient anxiety, fear of unnecessary and expensive tests, procedures that might follow from a genomic result, and privacy concerns.

Solutions to making precision medicine work include the following:

1. Establishing standards: Data is the bedrock of precision medicine collected from different people, through different sources, via different channels, and from different devices.
2. Robust business models: Precision medicine combines many realms: digital technologies, personal data, medical expertise, biochemistry, and healthcare provision.
3. Precision platforms: Effective digital platforms could be the answer to support the collaboration of all these different stakeholders

Other challenges and solutions with precision medicine

The other challenges in the implementation of precision medicine involve various factors as illustrated in Table 4.1 and literature [21]. These include the design of clinical trials and the low rate of matching patients to drugs in these trials, which ranges from 5% to 49% and is mostly in the 15% to 20% range and low outcome. The low rate of matching patients to drugs in these trials can be due to (1) enrollment of cancer patients with end-stage disease; (2) use of gene panels that yield little actionable targets; (3) timing and interpreting genomic test results; and (4) challenges involving the accessibility to targeted therapy and/or limited drug availability. Some solutions have been suggested such as the ones provided by trials

Table 4.1 Examples of precision medicine trials, design, and outcomes.

Clinical trial name	Cancer type	Trial type	Number of patients screened	% patients matched	Biomarkers assessed	Trial outcome	References
Bisgrove	Various treatment-refractory tumors	Prospective, navigational	86	77%	IHC, FISH, and microarray	27% of 66 matched pts had a PFS2/PFS1 ratio[a] HYPERLINK (CI, 17%–38%; P: .007).	[22]
BATTLE	Lung cancer	Prospective, adaptive, and randomized	255	Not available	11 biomarkers	8-week disease control rate, 46%	[23]
IMPACT, first cohort	Various treatment-refractory tumors	Registry type and navigational	1144	15%	PCR-based genomics, 9 genes	Matched vs. unmatched RR, 27% vs. 5% ($P < .0001$); TTF: median, 5.2 vs. 2.2 mos ($P < .0001$); OS: median, 13.4 vs. 9.0 mos (P .017)	[24]
I-SPY 1	Breast cancer	Neoadjuvant and correlative	237	Nonapplicable	IHC	pCR differs by subset	[25]
IMPACT, second cohort	Various treatment-refractory tumors	Registry type and navigational	1276	11%	PCR-based genomics, 18–50 genes	Matched vs. unmatched RR, 11.9% vs. 5% ($P < 0.0001$); PFS: median, 3.9 vs. 2.2 mos (P .001); OS: median, 11.4 vs. 8.6 mos (P .04)	[26]
Lung cancer mutation consortium	Lung cancer	Prospective	1537	17%	Multiplex genotyping, 10 genes	Improved OS with matched vs. unmatched therapy (P .006)	[27]
SAFIR01/UNICANCER	Breast cancer	Prospective	423	13%	Sanger sequencing (PIK3CA and AKT); aCGH	Matched group, ORR 9%	[28]
SHIVA	Various treatment-refractory tumors	Prospective and randomized	741	13%	Targeted NGS, ~50 genes	PFS not improved with matched therapy (P .41)	[29]

Study	Tumor type	Study design	N	Percentage	Technology	Outcome	Ref
PREDICT	Various treatment-refractory tumors	Registry type	347	25%	NGS, 182 or 236 genes	Matched vs. unmatched; higher rates of SD ≥ 6 months/PR/CR (P. 02) and PFS (P < .04); higher matching scores correlated with better OS: 15.7 vs. 10.6 mos (P. 04)	[30]
MD Anderson Personalized Cancer Therapy Initiative	Various treatment-refractory tumors	Prospective and navigational	500	24%	NGS, 236 genes	Higher matching scores correlated with higher rates of SD ≥ 6 months/PR/CR (P. 024), TTF (P. 0003), and OS (P. 05)	[31]
BATTLE-2	Lung cancer	Prospective, adaptive, and randomized	334	Nonapplicable	ALK, FISH, EGFR, and KRAS Sanger sequencing	KRAS alterations: longer PFS without erlotinib (P. 04); KRAS wild-type tumors: longer OS on erlotinib (P. 03)	[32]
IMPACT/COMPACT	Various treatment-refractory tumors	Prospective	1893	5%	Hot spot panel, 23 genes	Matched vs. unmatched. Higher ORR: 19% vs. 9%, (P. 026)	[33]
I-SPY 2	Breast cancer	Phase 2 adaptive design, neoadjuvant	Non-applicable	Nonapplicable	IHC, Mammaprint	Improved pCR rates in two study arms with drug addition: HER2+, HR (−): neratinib plus standard therapy (N: 115) vs. standard therapy (N: 78): 56% vs. 33%. Triple-negative: veliparib plus carboplatin (N: 72) with standard therapy vs. standard therapy (N: 44): 51% vs. 26%	[34]
IMPACT, third cohort	Various treatment-refractory tumors	Registry type and navigational	1436	27%	PCR-based genomics and NGS, 11–182 genes	Matched vs. unmatched. Higher rates of ORR (P. 0099), TTF (P. 0015), and OS (P. 04)	[35]
MOSCATO	Various treatment-refractory tumors	Prospective	1035	19%	Targeted NGS, 40–75 genes; aCGH; RNAseq	PFS2/PFS1 ratio[a] was >1.3 in 33% (63/193) of patients	[36]

(Continued)

Table 4.1 (Continued)

Clinical trial name	Cancer type	Trial type	Number of patients screened	% patients matched	Biomarkers assessed	Trial outcome	References
MyPathway	Various treatment-refractory tumors	Prospective, Phase 2 basket	251	Not available	Genomic testing via any CLIA lab	Matched patients; ORR: All, 23%; HER2-altered, 38%; BRAF-altered, 43%	[37]
Profiler	Various treatment-refractory tumors	Prospective	2579	6%	NGS, 69 genes	RR: 13% (23 of 182 treated)	[38]
Lung Cancer Mutation Consortium II	Lung cancer	Prospective	904	12%	NGS, minimum of 14 genes	Improved survival with matched therapy ($P < 0.001$)	[39]
I-PREDICT	Various treatment-refractory tumors	Prospective and navigational	149	49%	NGS, 315 genes; ctDNA; PDL1 IHC	Higher matching scores correlated with increased rates of SD ≥ 6 months/PR/CR: 50% vs. 22.4% ($P. 028$), PFS ($P. 0004$), and OS ($P. 038$)	[40]
WINTHER	Various treatment-refractory tumors	Prospective and navigational	303	35%	NGS, 236 genes; transcriptomics	Higher matching scores correlated with longer PFS ($P. 005$) and OS ($P. 03$)	[41]
VICTORY	Gastric cancer	Prospective	772	14%	NGS, IHC, PDL1, MMR, and EBV status	Improved PFS and OS with matched vs. unmatched therapy ($P < .0001$)	[40]

The major challenges in designing trials for precision medicine involve various factors, including the low rate of matching patients to drugs in these trials, which ranges from 5% to 49% and is mostly in the 15% to 20% range and low outcome. *aCGH*, Array comparative genomic hybridization; *ASCO*, American Society of Clinical Oncology; *CLIA*, clinical laboratory improvement amendment; *cDNA MA*, cDNA microarray; *CGP*, comprehensive genomic profiling; *CR*, complete remission; *ctDNA*, circulating tumor DNA; *FISH*, fluorescence in situ hybridization; *IHC*, immunohistochemistry; *mos*, months; *NGS*, next-generation sequencing; *ORR*, overall response rate; *OS*, overall survival; *pCR*, pathological complete response; *PCR*, polymerase chain reaction; *PR*, partial remission; *PFS*, progression-free survival; *pts*, patients; *RR*, response rate; *RRPA*, reverse phase protein array; *SD*, stable disease; *TTF*, time to treatment failure.

[a]PFS2/PFS1 ratio is defined by the PFS on trial versus the PFS on the therapy immediately preceding the trial; in general, PFS is shorter with every subsequent therapy.

Source: Adapted from Tsimberidou AM, Fountzilas E, Nikanjam M, Kurzrock R. Review of precision cancer medicine: evolution of the treatment paradigm. Cancer Treat Rev. 2020;86:102019.

(e.g., I-PREDICT 12) (matching rate, 49%). These suggestions include (1) use of clinical trial navigators and medication acquisition specialists; (2) use of a more comprehensive NGS panel with >200 genes; (3) creation of electronic molecular tumor boards immediately upon physician request; and (4) use of biomarkers to match patients to therapy.

As discussed in the literature [21], there are other challenges in the application of precision medicine, including (1) different responses to matched targeted therapy due to the histology and/or genomic comolecular alterations. In this case, specific genomic biomarkers can be predictive within specific tumor histology [42,43]; (2) heterogeneity between primary tumor and metastatic sites and evolution of genomic molecular signatures. Molecular profiling of tumor tissue obtained from a single lesion may not always represent the systemic disease [44,45]. Targeted treatments can lead to the emergence of resistant clones and novel molecular alterations driving disease progression [46,47]; (3) need to screen large numbers of patients to identify specific genomic defects (e.g., NTRK fusions) [42,43,48] using a comprehensive panel of genomics tests; (4) differences in the metabolism and adverse effects of study drugs in different ethnic groups; (5) discordance between assays from different diagnostic tests; (6) inability to access to targeted therapy drugs; and (7) stringent eligibility criteria that may exclude many patients with real-world comorbidities. Approximately 3%−5% of patients with cancer are enrolled in clinical trials, and accrual is limited by the stringent eligibility criteria and limited access to drugs [49].

From the Cancer Genome Atlas project to targeted therapy

The use of genetic and genomic information to develop personalized disease prevention strategies is now well established in the scientific community but not yet widely adopted in clinical practice.

As discussed in the literature [50], with the undertaking of the Cancer Genome Atlas (TCGA) project (a joint effort of the National Cancer Institute (NCI) and the National Human Genome Research Institute (NHGRI) in 2005 to map the human cancer genome [51,52], the TCGA project has identified new oncogenic point mutations, fusions, and variants that have an impact on therapeuty and on the clinical course of cancer patients. This effort led to the first steps toward precision medicine in oncology with the goal that cancer patients get the right treatment at the

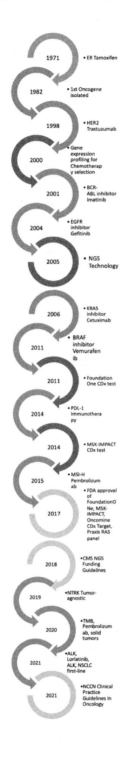

right dose at the right time, with minimum adverse events and maximum efficacy [53].

Although the use of tamoxifen in ER-positive breast cancer [54] could be the first example of precision oncology, more relevant precision oncology drugs approved against molecular targets were trastuzumab (for HER2 + breast cancer in 1998) and imatinib [for BCR/ABL-positive chronic myelogenous leukemia (CML) in 2001].

Another example is that certain mutations in lung cancer are being clinically targeted, such as those in EGFR and ALK genes. However, one of the most frequently mutated genes in NSCLC is the KRAS. Approximately 25% of all types of NSCLC tumors contain KRAS mutations. These mutations are indicative of poor prognosis and show negative response to standard chemotherapy. Furthermore, tumors harboring KRAS mutations are unlikely to respond to currently available targeted treatments such as tyrosine kinase inhibitors. Therefore, there is a definitive, urgent need to generate new targeted therapy approaches for KRAS mutations in which there are targeted therapies in development.

Recently, there have been advances in targeting mutant KRAS, including KRAS (G12C)-specific inhibitors such as AMG510 (sotorasib, $KRAS^{G12C}$), MRTX849 (adagrasib, $KRAS^{G12C}$), and $KRAS^{G12C}$ inhibitor MRTX1133 ($KRAS^{G12D}$), which are in development and some have obtained encouraging results in clinical trials [55,56] and notably, AMG510 was the first $KRAS^{G12C}$-targeted drug approved in 2021 [55].

In the ensuing years, the number of actionable genomic alterations that have a matching targeted therapy has increased steadily. These genomic alterations include both single genes (e.g., BRAF and ALK), or composite genetic signatures (e.g., mismatch repair or homologous recombination deficiency). Fig. 4.3 illustrates a timeline of the examples of precision oncology highlights (Fig. 4.4).

Figure 4.4 Timeline illustrating the major precision oncology highlights, including therapeutic landmarks and their molecular targets (in green), most relevant diagnostic technologies (in blue), and regulatory landmarks (in yellow). NCCN, National Comprehensive Cancer Network. *Adapted from Mateo J, Steuten L, Aftimos P, Andre F, Davies M, Garralda E, et al. Delivering precision oncology to patients with cancer. Nat Med 2022;28:658−65; Colomer R, Mondejar R, Romero-Laorden N, Alfranca A, Sanchez-Madrid F, Quintela-Fandino M. When should we order a next generation sequencing test in a patient with cancer? EClinicalMedicine 2020;25:100487.*

Targeted cancer therapy

The druggable genome where genes and gene products are known or predicted to interact with orally bioavailable compounds has been the focus of drug discovery and development [5,57,58].

While many of the druggable targets have a well-defined protein structure that can be bound by small molecules, targeting many other potential druggable proteins that play important biological functions has been challenging due to the lack of potential competitive binding sites for small drugs.

Several novel therapeutic agents, including kinase inhibitors, antibody-drug conjugates, cytotoxic drugs, hormone therapy, immunotherapy, etc., have been developed [57]. In recent years, many targeted therapies have been approved by the FDA for the treatment of different types of cancer with specific genetic, genomic, or epigenetic alterations paired with specific targeted therapies resulting in a positive outcome for many cancer patients (Fig. 4.5).

For instance, anti-CD20 monoclonal antibody rituximab was approved for the treatment of low-grade B cell lymphoma in 1997, and thereafter, the anti-HER2 monoclonal antibody trastuzumab was approved in 1998 for the treatment of HER2 overexpressing breast cancer.

Recently, monoclonal anti-PD-1 blocking antibody pembrolizumab was approved in 2014 for the treatment of melanoma, lung cancer, head and neck cancer, lymphoma, and several other indications.

More recently, the anti-CD19 cytolytic monoclonal antibody tafasitamab-cxix was approved for the treatment of diffuse large B-cell lymphoma in 2020. ABECMA, a genetically modified personalized immune cell therapy, to recognize and attack multiple myeloma cells that express B-cell maturation antigen (BCMA) was approved in 2021 for the treatment of adult patients with relapsed or refractory multiple myeloma. Breyanzi, a CD19-directed chimeric antigen receptor (CAR) T-cell therapy, was approved in 2021 for the treatment of adult patients with relapsed or refractory large B-cell lymphoma. CARVYKTI, a BCMA-directed, genetically modified autologous T-cell immunotherapy, involves reprogramming a patient's own T cells with a transgene encoding a CAR that identifies and eliminates cells that express BCMA was approved in 2021 for the treatment of adult patients with relapsed or refractory multiple myeloma. As of mid-2022, CA-31 biologics have been approved by the FDA for the treatment of different types of cancers (Drugs@FDA data, October 2020) [5,57].

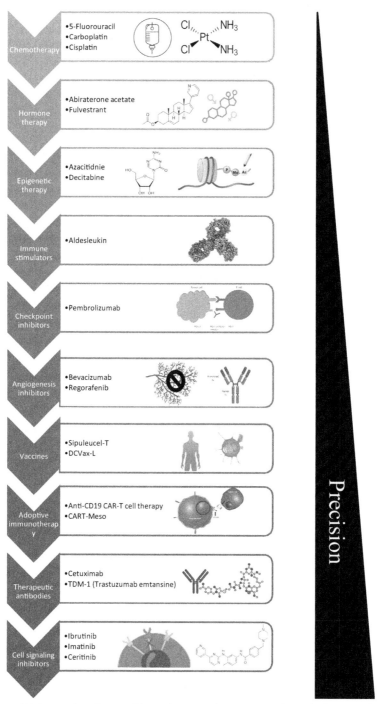

Figure 4.5 Increasing the precision of cancer therapeutics through the use of more targeted therapies. The illustration of the type of targeted cancer therapy with corresponding examples of drugs with increasing precision [59]. *Adapted from Ravoori P., What is precision medicine? Cancer Res Catal - Am Assoc Cancer Res 2015.*

https://www.fda.gov/vaccines-blood-biologics/development-approval-process-cber/2021-biological-license-application-approvals

Protein kinases are key components of intracellular signaling pathways involved in various cellular functions. Often those pathways are dysregulated in cancer; thus, inhibiting protein activity with small-molecule inhibitors is another way to successfully execute the targeted therapy strategy. Examples of these types of protein kinase-targeted therapy include protein kinase inhibitors (e.g., imatinib). Imatinib was developed using a rational drug design to target the fusion protein BCR-ABL driving CML and was approved by the FDA in 2001 [20]. Following this, several other small-molecule kinase inhibitors have been developed (Fig. 4.5) [60−62].

Most kinase inhibitors compete with ATP-competitive inhibitors. To add this issue, the next-generation non-ATP-competitive allosteric kinase inhibitors such as mTOR inhibitors temsirolimus and everolimus, followed by MEK inhibitors trametinib, cobimetinib, binimetinib, and selumetinib that bind kinases outside of the ATP pocket, were developed and approved by the FDA for the treatment of different cancer types. Those non-ATP-competitive allosteric kinase inhibitors are more specific than ATP-competitive inhibitors. Indeed, they target protein domains with less homology within the kinases than the ATP pocket, leading to decreased off-target effects and toxicity profiles [63].

In addition, small-molecule covalent kinase inhibitors such as the BTK inhibitor ibrutinib and afatinib, which target EGFR, HER2, HER4, and some EGFR mutants resistant to the first-generation EGFR inhibitors, were also developed and approved by the FDA in 2013.

These small-molecule covalent kinase inhibitors bind to their target with covalent bonds, which leads to superior binding affinity, potency, specificity, prolonged pharmacodynamics, and reduced off-target effects and toxicity [64]. The strategy for the development of these small-molecule covalent kinase inhibitors has better PK, greater target specificity, fewer off-target effects, and to a certain extent, overcoming drug resistance.

Challenges involving targeted therapies

Despite the recent successes with targeted therapies, there have been some limitations. Tumors are heterogeneous; thus, cells escape treatments mounting drug resistance. In addition to genetics, other factors such as tumor burden heterogeneity, the type of immune response, the tumor microenvironment (TME), and environmental factors (like the type of work and diet) play a significant role in making the tumor resistant to therapy [65]. Several attempts have been made to address resistance to targeted therapy. For instance, combinatorial therapy may be a good option to tackle drug resistance [66,67]. Another approach is to use a targeted therapy in combination with one or more chemotherapies. A good example is trastuzumab (HER2/neu receptor inhibitor) used in combination with traditional chemotherapy to treat patients with metastatic breast cancer with HER2/neu overexpression.

Mutations that lead to RAS gene activation play important roles in the oncogenic transformation of almost one-quarter of all human cancers [68−71]. Previous studies have shown that targeting RAS mutants directly by genetic knockdown or silencing approaches has been effective in inhibiting RAS-driven cancers [72,73]. Recent clinical trials of G12C allele-specific covalent inhibitors exhibited high dose tolerance and effective antitumor effect in certain patients, indicating that selective inhibition of KRAS^{G12C} mutant cancers is a potential therapeutic strategy to be followed [68,74,75]. Likewise, a monobody, called 12VC1, that recognizes the active state of both KRAS^{G12V} and KRAS^{G12C} up to 400 times more tightly than wild-type KRAS has been reported [70]. Recently, there have been advances in directly targeted drugs for KRAS, especially in KRAS^{G12C} inhibitors, such as AMG510 (sotorasib, KRAS^{G12C}), MRTX849 (adagrasib, KRAS^{G12C}), and MRTX1133 (KRAS^{G12D}), which are investigational, highly selective, noncovalent, and potent small molecule inhibitors in development, and some have obtained encouraging results in clinical trials [55,56]. Notably, AMG510 was the first KRAS^{G12C}-targeted drug to be approved for clinical use in 2021 [55]. Despite all those efforts, recent evidence indicates that even a successful inhibition of RAS in preclinical models might lead to drug resistance. A recent report has shown that loss of KRAS, albeit a genetic KO model, diminishes, but does not abolish the tumorigenic capacity of PDAC cells and BRAF and MYC partially rescue tumorigenesis of KRAS KO cells in immunocompetent mice [76].

A recent case report [77] has reported two patients, one targeted therapy, and different outcomes. Both of these patients had advanced cancer resistant to multiple lines of systemic therapy and received a PARP inhibitor [77]. First case involved a 71-year-old male patient diagnosed with localized prostate cancer resistant to multiple lines of systemic therapy and received a PARP inhibitor based on the molecular profiling by next-generation DNA sequencing with the Oncomine gene panel (143 cancer genes) of the patient's tumors alongside their healthy cells to distinguish sequence changes unique to the tumor. Case 2 involved a 62-year-old male patient underwent resection of a high-grade sarcoma involvement of one of three lymph nodes resistant to multiple lines of systemic therapy and received a PARP inhibitor based on the molecular profiling by next-generation DNA sequencing as above. In case 1, a somatically acquired nonsense mutation in 32% of the ATM gene copies were identified. In addition to the ATM mutation, amplifications of AR and MYC genes were found in the tumor sample, as well as several lower level amplifications of other genes. Analysis of the tumor RNA for gene fusions identified a chimeric transcript having the 5' end of the TMPRSS2 (androgen-sensitive) gene joined to the 3' end of ERG (transcription factor). In case 2, next-generation sequencing of DNA analysis revealed a somatically acquired frameshift mutation in 39% of the ATM gene copies and additional mutations in NRAS and TP53. The patient with prostate cancer had prolonged stable disease which is ongoing; however, the patient with sarcoma had disease progression and died soon after initiating therapy. Despite both patients' tumors having a damaging ATM mutation, the difference in outcomes questions the variable effects of targeted therapies in patients with destructive mutations in the same gene. Among several other reasons, including both patients having different tumor types, different tumor heterogeneity, different specific mutations in ATM, differences in comutations, and epigenetic regulation of gene expression might lead to disparate responses in these cases. As such, understanding the contribution of these and other factors to drug responses might help to address a major challenge to precision medicine. Nevertheless, the precision medicine strategy for cancer treatment is still evolving.

As a result, in addition to the suggestions provided earlier mentioned to improve the success of targeted therapies, there is a need to accelerate not only the adoption of genome-targeted approaches and promote novel genome-targeted approaches but also move away from a drug-centered approach to a patient-centered approach, more effective use of companion diagnostics, and adoption of innovative clinical trial designs to improve patient outcomes. The innovative clinical trial designs include tumor agnostic trials [48,78−80], basket [48], umbrella [81], platform, octopus, and master protocols [21,24,26,30,35]. Randomization has also evolved, with the emergence of Bayesian adaptation, which allows dynamic modifications of randomization based on small numbers of patients and real-time outcomes.

Biomarkers in clinical practice

Biomarkers have been one of the pillars of medicine since the time of Hippocrates and have become even more important in recent years. The use of HER2 expression levels as the indicator of treatment response for breast cancer was recognized as some of the markers indicative of a pathological state and disease. Later, ER, PR, Ki67, PDL-1, and TILs were added to the arsenal of biomarkers for breast cancer (Table 4.2). Another good example of the stratification need is KRAS mutation, which is present in 25% of all types of NSCLC and these patients have a poor prognosis. Of course, there are many more examples that can be quoted.

Table 4.2 Top selected examples of biomarkers by cancer type and years.

Cancer type	Representative biomarkers (2000−18)			Representative Biomarkers (2016−18[b])		
Brain/glioma/CNS	EGFR	MTOR	IDH1/2	CD4	BRAF	CD8
Breast cancer	HER2	ER[a]	PR[a]	Ki67	PD-L1	TILs
Colorectal cancer	KRAS	EGFR	BRAF	MSI	NRAS	CD8
Head and neck cancer	HPV	EGFR	p16	PD-L1	CD8	TILs
Leukemia	MRD	BCR-ABL1	FLT3	CD19	BTK	TP53
Lung	EGFR	ALK	PD-L1	ROS1	KRAS	PD1
Lymphoma	CD20	CD4	MRD	CD19	MYC	CD30
MDS[a]	HLA-A	MRD	FLT3	CD4	CD8	CD3
Melanoma	BRAF	CD8	TIL	PD-L1	PD1	CD4
Myeloma	MRD	HLA-A	PSA	LDH	IL-6	EGFR
Ovarian cancer	BRCA1	BRCA2	CD8	PD-L1	TILs	CD8
Pancreatic cancer	EGFR	KRAS	PSA	CD8	PD-L1	IFN
Prostate cancer	PSA	AR	CTC	ATM	BRCA2	BRCA1
Renal cancer	MTOR	VEGF	EGFR	PD-L1	CD8	PD1
Stomach cancer	HER2	KIT	EGFR	PD-L1	TILs	PD1
Pan-tumor	MSI^	NTRK^	TMB	FGFR		

Examples of selected biomarkers used in clinical trials that are specific to the mechanisms targeted by targeted drugs and immunotherapies.
[a]ER: Estrogen receptor; *MDS*, myelodysplastic syndromes; *PR*, progesterone receptor.
[b]Includes biomarkers not already included in the top 3 from 2000 to 2018. Biomarkers with therapies approved for pan-tumor use. Adapted from Aggregate Analysis of ClinicalTrials.gov (AACT) database (https://aact.ctti–clinicaltrials.org/).

Biomarkers may be the best way to address the decision-making process by asking, "Is there evidence that the drug is working in humans using the same mechanism of action defined by animal studies?" Can target engagement be demonstrated? Traditional clinical endpoints such as morbidity and mortality often require extended time frames and are difficult to evaluate. A fully validated surrogate marker can instead be used as the primary clinical endpoint for pivotal studies. The presence of a validated biomarker has underlined the relevance of stratifying patients in responders and nonresponders by using a biomarker. Identifying upfront responders can have a profound impact on the size and cost of a clinical trial. Eliminating nonresponders reduces the intersubject variability and the number of subjects needed to demonstrate efficacy, sparing the nonresponders from drug toxicity. Because of their tremendous value, the developments of targeted therapies and diagnostic tests are performed simultaneously to guide the safe and effective use of a corresponding targeted therapy, which is defined as companion diagnostics by the US FDA [82].

In recent years, companion diagnostics have become an integral part of PM [4,83]. Indeed, it has been demonstrated that almost 40% of drugs are ineffective in patients with a variety of conditions, including cancer. As a result, companion diagnostics can help to identify patients who might benefit from a given therapy. As of July 2022, there were ca. 145 cleared or approved oncology companion diagnostic biomarkers reported on the FDA portal.

(https://www.fda.gov/medical-devices/in-vitro-diagnostics/list-cleared-or-approved-companion-diagnostic-devices-in-vitro-and-imaging-tools).

Although many new biomarkers have been approved for new therapies, unfortunately, most patients do not have access to them [83]. Hampering the use of those biomarker assays includes slow laboratory test adoption, the lack of test reimbursement, and the deficiency, or delay in the inclusion of diagnostic testing to the clinical guidelines [83].

Early and late examples of precision medicine

Early examples

The following are some historical examples of personalized disease treatments. Many early examples of precision medicine were associated with genetically mediated pharmacokinetic aspects of drugs. Warfarin is a widely used blood thinning medication that, if not dosed properly, could cause a potentially life-threatening adverse drug reaction. Warfarin targets a particular gene, VKORC1, and is metabolized in part by the gene CYP2C9. Naturally occurring genetic variation in both the VKORC1 and CYP2C9 genes leads to variation in the pharmacodynamic and pharmacokinetic properties of warfarin. The FDA has recommended that before warfarin dosing the individual genotype has to be taken into consideration [84].

Here we show a couple of examples of drugs that should only be provided to individuals with a certain genetic profile. Primaquine (PQ) has been used to manage malaria with some success in parts of the world where malaria is endemic. However, military doctors working in the past with this drug observed that some of the soldiers they treated for malaria that were provided the drug became jaundiced and anemic and ultimately exhibited symptoms that would later be termed "acute hemolytic anemia (AHA)." Later on, it was shown that the individuals were exhibiting AHA after PQ administration carried variants in the G6PD gene [85]. Another example is imatinib, which is used to treat CML. Imatinib inhibits an enzyme, tyrosine kinase, that is increased by the formation of a fusion of two genomic regions, one encompassing the Abelson proto-oncogene (ABL) and the other the breakpoint cluster region (BCR) [86]. This resulted in significant clinical remissions, leading to its approval in 2001 by the FDA [20]. Finally, another good example in solid tumor oncology was testing the tumor specimens of patients for *HER2* gene amplification (by fluorescent in situ hybridization) or protein overexpression (by immunohistochemistry). In this case, *HER2*-positive tumors predicted the patient's response to HER2-targeted trastuzumab therapies. Conversely, *HER2*-negative tumors did not benefit from this therapy.

More recent examples involving mutation-specific therapies

Instead of developing a drug and then identifying factors that mitigate its efficacy or side effects through observational studies, like the case with warfarin, PQ, and imatinib, there have been attempts to identify genetic profiles possessed by patients and then craft therapies that uniquely target those profiles. For example, the drug ivacaftor was designed to treat individuals with cystic fibrosis (CF) who have very specific pathogenic mutations in the gene cystic fibrosis transmembrane conductance regulator (CFTR) anion channel [87]. The CFTR gene has many functions, but one set of functions is dictated by a "gate-like" structure in the CFTR gene's encoded protein that can open and close to control the movement of salts in and out of cells. If the CFTR gene is dysfunctional, then the gate is closed, causing a build-up of mucus in the lungs. Ivacaftor is designed to open the gate for longer periods in the presence of certain mutations that tend to cause the gate to be closed.

An additional example involves the emerging set of cancer treatments known as immunotherapies [88]. Although there are many types of immunotherapies, all of them seek to prime or trigger an individual's own immune system to attack cancer. One type of immunotherapy exploits potentially unique sets of genetic alterations that arise in a patient's tumor cells, known as "neo-antigens," which are often capable of raising an immune response if recognized properly by the host's immune system. Essentially, this type of immunotherapy works by harvesting cells from a patient that mediate patient's immune reactions, such as T cells, modifying those cells to specifically recognize and target the neo-antigens found to be present in the patient's tumor. These modified cells are then put back in the patient's body so these cells can attack the tumor cells giving off the neo-antigen signals. Cytotoxic T-cell therapies like this, as well as immunotherapies in general, have had notable successes, but can be very patient-specific for two reasons: First, the neo-antigen profile of a patient might be very unique, such that cytotoxic T cells made to recognize and attack a specific set of neo-antigens will not work in someone whose tumor does not have those neo-antigens. Second, if "autologous" constructs are used, the patient's own T cells are modified, hence, not likely to work in another patient, although attempts to make "allogeneic" constructs in which one individual's T cells are modified and introduced into another patient's body are being pursued aggressively.

Third example is provided by lung cancer, the leading cause of mortality among all cancer types worldwide. Certain mutations are being clinically targeted, such as those in EGFR and ALK genes. However, one of the most frequently mutated genes in NSCLC is the Kirsten rat sarcoma viral oncogene homolog (KRAS), which is currently not targetable. Approximately 25% of all types of NSCLC tumors contain KRAS mutations, which remain as an undruggable challenge. These mutations are indicative of poor prognosis and show a negative response to standard chemotherapy. Furthermore, tumors harboring KRAS mutations are unlikely to respond to currently available targeted treatments such as tyrosine kinase inhibitors. Therefore, there is a definitive, urgent need to generate new targeted therapy approaches for KRAS mutations.

Conclusions and future perspectives

Rapid development and adaptation of various high-throughput "omics" platforms, including molecular profiling technologies, computational biology, bioinformatics, and more recently machine learning and AI, helped to improve precision medicine efforts by analyzing large data sets generated from these efforts. The development of cutting-edge technologies, machine learning, and AI-based big data platforms have the potential to revolutionize the field of medicine and allow a high volume of data to be analyzed quickly. While this poses unprecedented challenges in data storage, processing, exchange, and curation, it will ultimately provide the scientific community a better understanding of disease biology by going beyond single genetic mutations or alterations to other molecular alterations to construct causal network models, molecular signatures, and actionable targets involved in disease initiation, progression, and response to therapies.

As a result, AI and machine learning will improve the traditional symptom-driven practice of medicine allowing earlier interventions using advanced diagnostics and tailoring better and economically personalized treatments.

In conclusion, to implement precision medicine strategy, we need to improve omics testing procedures, develop preventive and therapeutic strategies, and build a library of information about how to use omics in health care.

Acknowledgments

AAS and CC wrote the manuscript and agree with the manuscript's results and conclusions.

Ethics approval and consent to participate

This article does not contain any studies with human participants or animals performed by any of the authors. Informed consent was not required for the preparation of this review article as it used secondary sources only.

Consent for publication

The authors grant the publisher the sole and exclusive license of the full copyright in the contribution, which license the publisher hereby accepts.

Competing interest

The authors declare that he has no competing interests.

Abbreviations

aCGH	array comparative genomic hybridization
AI	artificial intelligence
ASCO	American Society of Clinical Oncology
BM	biomarker
CAR-T	chimeric antigen receptor T cells
cDNA MA	cDNA microarray
CDx	companion diagnostics
CGP	comprehensive genomic profiling
CLIA	clinical laboratory improvement amendment
CML	chronic myeloid leukemia
CR	complete remission
ctDNA	circulating tumor DNA
CTLA-4	cytotoxic T-lymphocyte-associated protein 4
EBV	Epstein−Barr virus
ER	Estrogen receptor
FDA	Food and Drug Administration
FISH	fluorescence in situ hybridization
IHC	immunohistochemistry
LBx	liquid biopsies
MDS	myelodysplastic syndromes
MMR	mismatch repair
MOS	months
NCI	National Cancer Institute
NHGRI	the National Human Genome Research Institute
NGS	next-generation sequencing
NSCLC	nonsmall cell lung cancer
ORR	overall response rate
OS	overall survival
pCR	pathological complete response
PCR	polymerase chain reaction
PD-1	programmed cell death protein 1
PDL-1	programmed death-ligand 1
PFS	progression-free survival
PD	pharmacodynamics
PK	pharmacokinetics
PM	precision medicine
PR	partial remission
PR	progesterone receptor
PTS	patients
RR	response rate
RRPA	reverse phase protein array

SD	stable disease
TCGA	the Cancer Genome Atlas
TME	tumor microenvironment
TTF	time to treatment failure

References

[1] Demirkaya E, Arici ZS, Romano M, Berard RA, Aksentijevich I. Current state of precision medicine in primary systemic vasculitides. Front Immunol 2019;10:2813.

[2] Berman JJ. - Precision data. In: Berman JJ, editor. Precision medicine and the reinvention of human disease. Academic Press; 2018. p. 263–326.

[3] Waring MJ, Arrowsmith J, Leach AR, Leeson PD, Mandrell S, Owen RM, et al. An analysis of the attrition of drug candidates from four major pharmaceutical companies. Nat Rev Drug Discov 2015;14:475–86.

[4] Seyhan AA, Carini C. Are innovation and new technologies in precision medicine paving a new era in patients centric care? J Transl Med 2019;17:114.

[5] Seyhan AA. The current state of precision medicine and targeted-cancer therapies: where are we? In: Scotti MT, Bellera CL, editors. Drug target selection and validation computer-aided drug discovery and design, 1. 2022. p. 119–200.

[6] Seyhan AA. Lost in translation: the valley of death across preclinical and clinical divide — identification of problems and overcoming obstacles. Transl Med Commun 2019;4:18.

[7] Kosorok MR, Laber EB. Precision medicine. Annu Rev Stat Appl 2019;6:263–86.

[8] Seyhan AA. In: Claudio Carini MF, Alain van Gool, editors. Lost translation — Chall use Anim Model Transl Res. Chapman and Hall/CRC; 2019. p. 36.

[9] Mateo J, Steuten L, Aftimos P, Andre F, Davies M, Garralda E, et al. Delivering precision oncology to patients with cancer. Nat Med 2022;28:658–65.

[10] Mosele F, Remon J, Mateo J, Westphalen CB, Barlesi F, Lolkema MP, et al. Recommendations for the use of next-generation sequencing (NGS) for patients with metastatic cancers: a report from the ESMO Precision Medicine Working Group. Ann Oncol 2020;31:1491–505.

[11] Krzyszczyk P, Acevedo A, Davidoff EJ, Timmins LM, Marrero-Berrios I, Patel M, et al. The growing role of precision and personalized medicine for cancer treatment. Technol (Singap World Sci) 2018;6:79–100.

[12] Lee YT, Tan YJ, Oon CE. Molecular targeted therapy: treating cancer with specificity. Eur J Pharmacol 2018;834:188–96.

[13] Perez-Herrero E, Fernandez-Medarde A. Advanced targeted therapies in cancer: drug nanocarriers, the future of chemotherapy. Eur J Pharm Biopharm 2015;93:52–79.

[14] Wang W, Sun Q. Novel targeted drugs approved by the NMPA and FDA in 2019. Signal Transduct Target Ther 2020;5:65.

[15] Malone ER, Oliva M, Sabatini PJB, Stockley TL, Siu LL. Molecular profiling for precision cancer therapies. Genome Med 2020;12:8.

[16] Ren R. Mechanisms of BCR-ABL in the pathogenesis of chronic myelogenous leukaemia. Nat Rev Cancer 2005;5:172–83.

[17] Cocco E, Scaltriti M, Drilon A. NTRK fusion-positive cancers and TRK inhibitor therapy. Nat Rev Clin Oncol 2018;15:731–47.

[18] Sokolenko AP, Imyanitov EN. Molecular diagnostics in clinical oncology. Front Mol Biosci 2018;5:76.

[19] Hawgood S, Hook-Barnard IG, O'Brien TC, Yamamoto KR. Precision medicine: Beyond the inflection point. Sci Transl Med 2015;7:300ps17.

[20] Schwartzberg L, Kim ES, Liu D, Schrag D. Precision oncology: who, how, what, when, and when not? American Society of Clinical Oncology Educational Book. 2017. p. 160–9.

[21] Tsimberidou AM, Fountzilas E, Nikanjam M, Kurzrock R. Review of precision cancer medicine: Evolution of the treatment paradigm. Cancer Treat Rev 2020; 86:102019.

[22] Von Hoff DD, Stephenson Jr. JJ, Rosen P, Loesch DM, Borad MJ, Anthony S, et al. Pilot study using molecular profiling of patients' tumors to find potential targets and select treatments for their refractory cancers. J Clin Oncol 2010;28:4877–83.

[23] Kim ES, Herbst RS, Wistuba II, Lee JJ, Blumenschein Jr. GR, Tsao A, et al. The BATTLE trial: personalizing therapy for lung cancer. Cancer Discov 2011;1:44–53.

[24] Tsimberidou AM, Iskander NG, Hong DS, Wheler JJ, Falchook GS, Fu S, et al. Personalized medicine in a phase I clinical trials program: the MD Anderson Cancer Center initiative. Clin Cancer Res 2012;18:6373–83.

[25] Esserman LJ, Berry DA, DeMichele A, Carey L, Davis SE, Buxton M, et al. Pathologic complete response predicts recurrence-free survival more effectively by cancer subset: results from the I-SPY 1 TRIAL–CALGB 150007/150012, ACRIN 6657. J Clin Oncol 2012;30:3242–9.

[26] Tsimberidou AM, Wen S, Hong DS, Wheler JJ, Falchook GS, Fu S, et al. Personalized medicine for patients with advanced cancer in the phase I program at MD Anderson: validation and landmark analyses. Clin Cancer Res 2014;20: 4827–36.

[27] Kris MG, Johnson BE, Berry LD, Kwiatkowski DJ, Iafrate AJ, Wistuba II, et al. Using multiplexed assays of oncogenic drivers in lung cancers to select targeted drugs. JAMA. 2014;311:1998–2006.

[28] Andre F, Bachelot T, Commo F, Campone M, Arnedos M, Dieras V, et al. Comparative genomic hybridisation array and DNA sequencing to direct treatment of metastatic breast cancer: a multicentre, prospective trial (SAFIR01/UNICANCER). Lancet Oncol 2014;15:267–74.

[29] Le Tourneau C, Delord JP, Goncalves A, Gavoille C, Dubot C, Isambert N, et al. Molecularly targeted therapy based on tumour molecular profiling versus conventional therapy for advanced cancer (SHIVA): a multicentre, open-label, proof-of-concept, randomised, controlled phase 2 trial. Lancet Oncol 2015;16:1324–34.

[30] Schwaederle M, Parker BA, Schwab RB, Daniels GA, Piccioni DE, Kesari S, et al. Precision oncology: the UC San Diego Moores Cancer Center PREDICT Experience. Mol Cancer Ther 2016;15:743–52.

[31] Wheler JJ, Janku F, Naing A, Li Y, Stephen B, Zinner R, et al. Cancer therapy directed by comprehensive genomic profiling: a single center study. Cancer Res 2016;76:3690–701.

[32] Papadimitrakopoulou V, Lee JJ, Wistuba II, Tsao AS, Fossella FV, Kalhor N, et al. The BATTLE-2 study: a biomarker-integrated targeted therapy study in previously treated patients with advanced non-small-cell lung cancer. J Clin Oncol 2016; 34:3638–47.

[33] Stockley TL, Oza AM, Berman HK, Leighl NB, Knox JJ, Shepherd FA, et al. Molecular profiling of advanced solid tumors and patient outcomes with genotype-matched clinical trials: the Princess Margaret IMPACT/COMPACT trial. Genome Med 2016;8:109.

[34] Park JW, Liu MC, Yee D, Yau C, van 't Veer LJ, Symmans WF, et al. Adaptive randomization of neratinib in early breast cancer. N Engl J Med 2016;375:11–22.

[35] Tsimberidou AM, Hong DS, Ye Y, Cartwright C, Wheler JJ, Falchook GS, et al. Initiative for molecular profiling and advanced cancer therapy (IMPACT): an MD Anderson Precision Medicine Study. JCO Precis Oncol. 2017;2017.

[36] Massard C, Michiels S, Ferte C, Le Deley MC, Lacroix L, Hollebecque A, et al. High-throughput genomics and clinical outcome in hard-to-treat advanced cancers: results MOSCATO 01 trial. Cancer Discov 2017;7:586−95.

[37] Hainsworth JD, Meric-Bernstam F, Swanton C, Hurwitz H, Spigel DR, Sweeney C, et al. Targeted therapy for advanced solid tumors on the basis of molecular profiles: results from MyPathway, an open-label, phase IIa Multiple Basket Study. J Clin Oncol 2018;36:536−42.

[38] Tredan O, Wang Q, Pissaloux D, Cassier P, de la Fouchardiere A, Fayette J, et al. Molecular screening program to select molecular-based recommended therapies for metastatic cancer patients: analysis from the ProfiLER trial. Ann Oncol 2019;30:757−65.

[39] Aisner DL, Sholl LM, Berry LD, Rossi MR, Chen H, Fujimoto J, et al. The impact of smoking and TP53 mutations in lung adenocarcinoma patients with targetable mutations—the Lung Cancer Mutation Consortium (LCMC2). Clin Cancer Res 2018;24:1038−47.

[40] Sicklick JK, Kato S, Okamura R, Schwaederle M, Hahn ME, Williams CB, et al. Molecular profiling of cancer patients enables personalized combination therapy: the I-PREDICT study. Nat Med 2019;25:744−50.

[41] Rodon J, Soria JC, Berger R, Miller WH, Rubin E, Kugel A, et al. Genomic and transcriptomic profiling expands precision cancer medicine: the WINTHER trial. Nat Med 2019;25:751−8.

[42] Hyman DM, Puzanov I, Subbiah V, Faris JE, Chau I, Blay JY, et al. Vemurafenib in multiple nonmelanoma cancers with BRAF V600 mutations. N Engl J Med 2015; 373:726−36.

[43] Ross JS, Ali SM, Fasan O, Block J, Pal S, Elvin JA, et al. ALK fusions in a wide variety of tumor types respond to anti-ALK targeted therapy. Oncologist. 2017; 22:1444−50.

[44] Lovly CM, Salama AK, Salgia R. Tumor heterogeneity and therapeutic resistance. Am Soc Clin Oncol Educ Book 2016;35:e585−93.

[45] Gerlinger M, Rowan AJ, Horswell S, Math M, Larkin J, Endesfelder D, et al. Intratumor heterogeneity and branched evolution revealed by multiregion sequencing. N Engl J Med 2012;366:883−92.

[46] Kobayashi S, Boggon TJ, Dayaram T, Janne PA, Kocher O, Meyerson M, et al. EGFR mutation and resistance of non-small-cell lung cancer to gefitinib. N Engl J Med 2005;352:786−92.

[47] Napolitano A, Vincenzi B. Secondary KIT mutations: the GIST of drug resistance and sensitivity. Br J Cancer 2019;120:577−8.

[48] Drilon A, Laetsch TW, Kummar S, DuBois SG, Lassen UN, Demetri GD, et al. Efficacy of larotrectinib in TRK fusion-positive cancers in adults and children. N Engl J Med 2018;378:731−9.

[49] Murthy VH, Krumholz HM, Gross CP. Participation in cancer clinical trials: race-, sex-, and age-based disparities. JAMA. 2004;291:2720−6.

[50] Colomer R, Mondejar R, Romero-Laorden N, Alfranca A, Sanchez-Madrid F, Quintela-Fandino M. When should we order a next generation sequencing test in a patient with cancer? EClinicalMedicine 2020;25:100487.

[51] Wheeler DA, Wang L. From human genome to cancer genome: the first decade. Genome Res 2013;23:1054−62.

[52] Green ED, Watson JD, Collins FS. Human genome project: twenty-five years of big biology. Nature. 2015;526:29−31.

[53] Mirnezami R, Nicholson J, Darzi A. Preparing for precision medicine. N Engl J Med 2012;366:489−91.

[54] Jordan VC. Fourteenth Gaddum Memorial Lecture. A current view of tamoxifen for the treatment and prevention of breast cancer. Br J Pharmacol 1993;110:507−17.

[55] Huang L, Guo Z, Wang F, Fu L. KRAS mutation: from undruggable to druggable in cancer. Signal Transduct Target Ther 2021;6:386.

[56] Wang X, Allen S, Blake JF, Bowcut V, Briere DM, Calinisan A, et al. Identification of MRTX1133, a noncovalent, potent, and selective KRAS(G12D) inhibitor. J Med Chem 2022;65:3123–33.

[57] Dupont CA, Riegel K, Pompaiah M, Juhl H, Rajalingam K. Druggable genome and precision medicine in cancer: current challenges. FEBS J 2021;.

[58] Hopkins AL, Groom CR. The druggable genome. Nat Rev Drug Discov 2002; 1:727–30.

[59] Ravoori P. What Is Precision Medicine? Cancer Research Catalyst - American Association for. Cancer Res 2015.

[60] Bhullar KS, Lagaron NO, McGowan EM, Parmar I, Jha A, Hubbard BP, et al. Kinase-targeted cancer therapies: progress, challenges and future directions. Mol Cancer 2018;17:48.

[61] Kannaiyan R, Mahadevan D. A comprehensive review of protein kinase inhibitors for cancer therapy. Expert Rev Anticancer Ther 2018;18:1249–70.

[62] Pottier C, Fresnais M, Gilon M, Jerusalem G, Longuespee R, Sounni NE. Tyrosine kinase inhibitors in cancer: breakthrough and challenges of targeted therapy. Cancers (Basel), 12. 2020.

[63] Panicker RC, Chattopadhaya S, Coyne AG, Srinivasan R. Allosteric small-molecule serine/threonine kinase inhibitors. Adv Exp Med Biol 2019;1163:253–78.

[64] Liu Q, Sabnis Y, Zhao Z, Zhang T, Buhrlage SJ, Jones LH, et al. Developing irreversible inhibitors of the protein kinase cysteinome. Chem Biol 2013;20:146–59.

[65] Vasan N, Baselga J, Hyman DM. A view on drug resistance in cancer. Nature. 2019;575:299–309.

[66] Eroglu Z, Ribas A. Combination therapy with BRAF and MEK inhibitors for melanoma: latest evidence and place in therapy. Ther Adv Med Oncol 2016;8:48–56.

[67] Tolcher AW, Peng W, Calvo E. Rational approaches for combination therapy strategies targeting the MAP kinase pathway in solid tumors. Mol Cancer Ther 2018; 17:3–16.

[68] Moore AR, Rosenberg SC, McCormick F, Malek S. RAS-targeted therapies: is the undruggable drugged? Nat Rev Drug Discov 2020;19:533–52.

[69] Cox AD, Fesik SW, Kimmelman AC, Luo J, Der CJ. Drugging the undruggable RAS: mission possible? Nat Rev Drug Discov 2014;13:828–51.

[70] Teng KW, Tsai ST, Hattori T, Fedele C, Koide A, Yang C, et al. Selective and noncovalent targeting of RAS mutants for inhibition and degradation. Nat Commun 2021;12:2656.

[71] Simanshu DK, Nissley DV, McCormick F. RAS proteins and their regulators in human disease. Cell. 2017;170:17–33.

[72] Acunzo M, Romano G, Nigita G, Veneziano D, Fattore L, Lagana A, et al. Selective targeting of point-mutated KRAS through artificial microRNAs. Proc Natl Acad Sci USA 2017;114:E4203–12.

[73] Sunaga N, Shames DS, Girard L, Peyton M, Larsen JE, Imai H, et al. Knockdown of oncogenic KRAS in non-small cell lung cancers suppresses tumor growth and sensitizes tumor cells to targeted therapy. Mol Cancer Ther 2011;10:336–46.

[74] Hallin J, Engstrom LD, Hargis L, Calinisan A, Aranda R, Briere DM, et al. The KRAS(G12C) inhibitor MRTX849 provides insight toward therapeutic susceptibility of KRAS-mutant cancers in mouse models and patients. Cancer Discov 2020; 10:54–71.

[75] Canon J, Rex K, Saiki AY, Mohr C, Cooke K, Bagal D, et al. The clinical KRAS (G12C) inhibitor AMG 510 drives anti-tumour immunity. Nature. 2019;575:217–23.

[76] Ischenko I, D'Amico S, Rao M, Li J, Hayman MJ, Powers S, et al. KRAS drives immune evasion in a genetic model of pancreatic cancer. Nat Commun 2021; 12:1482.

[77] Cecchini M, Walther Z, Sklar JL, Bindra RS, Petrylak DP, Eder JP, et al. Yale Cancer Center Precision Medicine Tumor Board: two patients, one targeted therapy, different outcomes. Lancet Oncol 2018;19:23−4.

[78] Hierro C, Matos I, Martin-Liberal J, Ochoa de Olza M, Garralda E. Agnostic-histology approval of new drugs in oncology: are we already there? Clin Cancer Res 2019;25:3210.

[79] Offin M, Liu D, Drilon A. Tumor-agnostic drug development. Am Soc Clin Oncol Educ Book 2018;38:184−7.

[80] Lacombe D, Burock S, Bogaerts J, Schoeffski P, Golfinopoulos V, Stupp R. The dream and reality of histology agnostic cancer clinical trials. Mol Oncol 2014; 8:1057−63.

[81] Herbst RS, Gandara DR, Hirsch FR, Redman MW, LeBlanc M, Mack PC, et al. Lung master protocol (lung-MAP)-A biomarker-driven protocol for accelerating development of therapies for squamous cell lung cancer: SWOG S1400. Clin Cancer Res 2015;21:1514−24.

[82] Al-Dewik NI, Younes SN, Essa MM, Pathak S, Qoronfleh MW. Making biomar-kers relevant to healthcare innovation and precision medicine. Processes 2022;10.

[83] Keeling P, Clark J, Finucane S. Challenges in the clinical implementation of preci-sion medicine companion diagnostics. Expert Rev Mol Diagn 2020;20:593−9.

[84] Lee MT, Klein TE. Pharmacogenetics of warfarin: challenges and opportunities. J Hum Genet 2013;58:334−8.

[85] Luzzatto L, Seneca E. G6PD deficiency: a classic example of pharmacogenetics with on-going clinical implications. Br J Haematol 2014;164:469−80.

[86] O'Dwyer ME, Druker BJ. Status of bcr-abl tyrosine kinase inhibitors in chronic myelogenous leukemia. Curr Opin Oncol 2000;12:594−7.

[87] Davis PB, Yasothan U, Kirkpatrick P. Ivacaftor. Nat Rev Drug Discov 2012; 11:349−50.

[88] Farkona S, Diamandis EP, Blasutig IM. Cancer immunotherapy: the beginning of the end of cancer? BMC Med 2016;14:73.

CHAPTER 5

Big Data and Health Analytics explained

Weronika Schary[1,2], Florian Brockmann[3], Jonathan Simantzik[1,2], Filip Paskali[1,2] and Matthias Kohl[1,2]

[1]Medical and Life Sciences Faculty, Furtwangen University, Villingen-Schwenningen, Germany
[2]Institute of Precision Medicine, Furtwangen University, Villingen-Schwenningen, Germany
[3]Division of Immunology, Department of Biology, University of Konstanz, Konstanz, Germany

Introduction

Big Data has become part of our everyday lives. Whether you're using a smartphone, a fitness tracking device, or social media, almost everything in our life generates data points that are being stored, analyzed, and evaluated often without us even thinking about it. The amount of data produced and stored in the Global Datasphere grows vastly every year (Fig. 5.1). Big Data has become a buzzword in modern society. Its meaning keeps evolving and adapting to technological and societal change. The beginning of the use of the term was back in the 1990s most likely in connection with data visualization [1,2]. At first, the term only described large and complex data sets that could not be analyzed, stored, or processed with the then conventional techniques. Since then, the concept of the term has broadened to include more than just the data set, extending to the methods and techniques used to produce the data as well as the approaches used to store, analyze, and process the data [3].

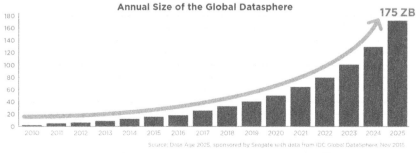

Figure 5.1 Annual size of the global datasphere in zettabytes (10^{21} bytes).

The New Era of Precision Medicine
DOI: https://doi.org/10.1016/B978-0-443-13963-5.00004-2

© 2024 Elsevier Inc.
All rights reserved.

115

Big Data in research

The well-established approach in research to focus on data samples rather than all the data was previously influenced by the cost of data collection and analysis, as well as overall data availability. Through the advances in digital technology, collecting, analyzing, and storing massive amounts of data has become faster and cheaper, leading research and especially bio-medical research into the Big Data era. But does more data equal more knowledge or does more data mean more noise? Big Data can look at all the data as a whole and provide new perspectives and new hypotheses. The key to standard analysis from data samples is the quality of data. For Big Data, the situation is a little different, because combining different data from different sources even with varying degrees of quality causes more heterogeneity, which can have a protective effect against measuring biases and produces good results depending on the size of the data sets. Simply put, the more data you have, the greater the likelihood of detect-ing and analyzing an effect. But phenomena and effects may change over time and constant new data acquisition might be required as well as adaptable models and re-evaluation of the data and the analysis. Big Data techniques differ from traditional statistical analysis in a way that they don't rely on fixed predefined research questions and hypotheses, but rather hypotheses, correlations, and connections arise from the analysis itself [4].

According to Hulsen et al. in 2019, Big Data research is rather a hypothesis generating than hypothesis-driven and enables research that will find unsuspected pathways that ask new questions with conventional methods still in place to test these new hypotheses [3].

Big Data has also become integral to many industries with sensors col-lecting data during the production process to improve a product, or infor-mation acquisition enabling pattern recognition to learn about customer satisfaction, consumer purchase behavior, and a whole lot more [5]. But Big Data also affects us much more closely and personally. Smart devices collect our personal data, and different apps try to improve our sleeping quality, our diet and activity levels, and our overall well-being.

In a much broader sense, the term Big Data in the media now also encompasses negative aspects of the collection, storage, and processing of large amounts of data, such as increasing surveillance by intelligence services, violation of customers' privacy rights by companies, increasing lack of transparency of data storage, industries trying to gain competitive advantages from existing data, microtargeting and personalized advertising, and so forth [6,7]. At the same time, Big Data also facilitates progress and comprehensive research in globally important domains such as ecology and environment, as well as medicine and health care.

In the medical field, the amount of health-related data is growing immensely (Fig. 5.2), and Big Data is nowadays closely linked to the term Health Analytics. Where the former is mainly concerned with the collection of large amounts of data from various sources, such as medical records, patient surveys, and health insurance claims, the latter refers to the accompanying analysis of this data to gain insights into patient health, disease trends, and healthcare delivery. Here, we will discuss the potential of Big Data and Health Analytics to improve healthcare outcomes and reduce costs, as well as the challenges associated with using those tools in health care, such as privacy and security concerns, and the need for skilled personnel to interpret the data.

Medicine has long strived to operate according to the principles of precision and individual therapy. Individual patient characteristics such as

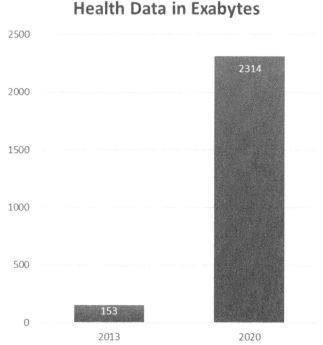

Figure 5.2 Growth in Health Data from 2013 to 2020 in exabytes (10^{18} bytes). *Adapted from Stanford Medicine 2017 Health Trends Report.*

age, gender, existing comorbidities, and patient preferences have influenced treatment decisions. However, what is new is the extent of the patient's molecular, genetic, and proteomic characteristics and even new sources of patient data, for example, from the so-called wearables that need to be considered for optimal therapy management of their diseases [2,8]. Especially through biomedical research and new techniques such as next-generation sequencing (NGS), whole genome sequencing (WGS) for sequencing DNA, and others that generate vast amounts of data in remarkably short periods of time, new molecular biological, genetic, or immunological characteristics are being considered (so-called biomarkers).

Using this data, which is also Big Data, in medicine and health care will further enable precision medicine, in which risk assessment for a specific disease, diagnosis of the disease and prognosis predicted with the individual patient's data, and interventions as well as therapy are tailored to the individual patient's needs. Furthermore, Big Data could detect global trends, identify at-risk groups, or find hidden links that can then be used to research the cause of a disease or to find a cure.

Precision medicine and personalized medicine

The terms precision medicine and personalized medicine are often used interchangeably, and the definition of the terms is constantly expanding. The work of Leroy Hood and the Institute for Systems Biology in Seattle has long been promoting the vision of P4 medicine [9−11]. P4 medicine, also known as predictive, preventive, personalized, and participatory medicine, is an emerging approach to health care that focuses on the prevention and early diagnosis of diseases, personalized treatments, and patient engagement. It is based on the idea that health care should be tailored to the individual, rather than relying on a one-size-fits-all approach. P4 medicine combines data from a variety of sources, such as genetics, lifestyle, environment, and medical history, to provide a more comprehensive view of a patient's health. The goal is to identify and treat diseases before they become serious and to provide personalized treatments that are tailored to the individual.

They report that patients continue to demonstrate an increasing interest to participate and manage their own health and health care. The rise of smart devices, fitness trackers, and other wearables enables digital tracking and reporting of symptoms and biomarkers and provides Big Data to power systems biology and medicine. Systems biology and medicine are oriented toward the big picture and intend to gather and connect all the pieces for a better understanding of complex processes in diseases and pathologies as well as to provide innovative approaches to health care, that is, prevention, prediction, diagnosis, and treatment of disease. Furthermore, they argue that the participation of patients in their own health care improves lifestyle decisions and produces the positive effect of lowering the incidence of chronic and complex diseases. They further explain that the digital revolution with the collection, integration, storage, and analysis of data and information facilitates this so-called P4 medicine. Systems medicine or P4 medicine will prove beneficial and even cost-effective by stratification of patients and diseases. For example, biological networks of disease will provide new drug targets for the pharmaceutical industry, and these drugs will be more effective for precise strata of patients and disease. Furthermore, drugs will be more cost-effective to develop and produce because smaller test populations will be required for these specific strata. Also, following the concept of P4 medicine, patients will be tested for specific biomarkers and stratified earlier and therefore interventions could take place at earlier stages of the disease, even before severe disease occurs, leading to possibly less hospitalization and more cost-effectiveness. They further argue that focusing on this concept could lead to a transformation of biomedical research and studies. Big Data analysis could soon replace averaging data from limited test cohorts, and control groups wouldn't need to be separate populations but could rather be extracted as subpopulations from analyzing massive heterogeneous data sets. Also, reclassification of diseases could occur, because diseases were classified, diagnosed, and treated on the basis of symptoms rather than underlying molecular and cellular mechanisms.

One research field that for a long time has generated the most data has been genetic and genomic research. Since the first draft of the human genome in the early 2000s and the rise of high-throughput technologies such as NGS, WGS, and more, the amount of sequencing data generated has grown exponentially. Over the years, the research has transformed from looking for protein-coding genes to explaining phenomena and especially diseases to zoom in even further and look for structural variations in genes or chromosomes such as copy-number variations, insertions, deletions, duplications, and even single-nucleotide polymorphisms (SNPs). Genome-wide association studies (GWASs) are performed to research these genetic variants, especially SNPs, and link them to traits or specific diseases. GWASs have made many important discoveries of genetic variants associated with complex human traits and diseases over the years, but still, the number of genetic variants discovered could not fully explain the observed heritability of the respective traits or diseases. This concept was coined as "missing heritability" [12,13].

For example, when the term "missing heritability" was introduced in 2008, about 40 genetic variants linked to height had been discovered explaining about 5% of the heritability of the trait. Several years later, about 700 associated genetic variants had been found explaining more than 20% of the heritability with further predictions that in the next few years, the number of discovered genetic variants associated with height could reach thousands and explain even more of the heritability [14]. New technologies and analysis techniques promised to solve the "missing heritability" mystery over the years only to discover that genetics and genomics are much more complicated and complex than the premise that one gene has only one function and leads to one specific trait. Moreover, findings suggest that pleiotropy exists for traits and complex diseases, which means that one gene or genetic variant influences two or more seemingly unrelated traits [14–17]. Another point that could be made in the case of the "missing heritability" is that GWASs were hypothesis-based studies and candidate genes and genetic variants were always studied in relation to a trait, but with the rise of Big Data GWAS and hypothesis-free approaches in recent years, many new genetic variants–trait associations were and will be discovered, possibly closing this gap [15,17].

Nevertheless, GWASs lead to great discoveries in the identification of genetic variants for type 2 diabetes, several autoimmune diseases as well as schizophrenia, and many others, providing new understanding for these diseases, new drug targets, and even opportunities for drug repurposing [14].

Through the rise of high-throughput sequencing and other omics research, such as transcriptomics, proteomics, metabolomics, etc., the availability of Big Data from biomedical research was gathered in databases such as UK Biobank [18], deCODE Iceland [19], KSCDC China [20], and many more. Many databases, especially the UK Biobank, are also a good example for data sharing enabling even more Big Data research.

Big Data for disease detection and pharmacovigilance

A completely different angle of using large amounts of data relies on data generated by individuals online amassing them and using vast amounts of computational power to infer dependencies from them. A widely known example of the attempt to detect disease outbreaks is Google Flu Trends

[21,22]. The idea was that search engine query data could be used to detect outbreaks of disease by tracking the frequency of searches related to symptoms of a particular disease. If there was a sudden increase in searches related to a particular symptom, it could be an indication of an outbreak of the disease. Additionally, search engine query contains geographical information and could therefore be used to track the spread of disease by analyzing the locations of the people making the searches. Google Flu Trends, launched by Google in 2008, worked by tracking the frequency of searches related to flu-like symptoms and comparing them to historical data. The project was able to accurately predict the spread of influenza in the United States with a lead time of one to two weeks. However, Google Flu Trends had some limitations. It was unable to accurately predict the spread of influenza in other countries, as the data was only collected from the United States. Additionally, the project was unable to accurately predict the severity of an outbreak, as it only tracked the frequency of searches related to flu-like symptoms. Another issue was that the system relied on a limited set of search terms, which meant that it could not accurately track the spread of the flu in areas where people used different terms to describe the same symptoms.

Another project was concerned with a novel approach to pharmacovigilance, which is the monitoring of the safety of drugs and other medical products [23]. The authors propose a web-scale pharmacovigilance system that uses natural language processing and machine learning techniques to identify signals from the crowd, such as online discussion forums, social media, and other sources of user-generated content. The system is designed to detect potential safety issues with drugs and other medical products and to alert healthcare professionals and regulatory agencies to potential safety concerns. The authors discuss the advantages of this approach, such as its scalability, its ability to detect signals from a variety of sources, and its potential to improve the safety of drugs and other medical products. The paper also provides an overview of the system architecture and implementation and discusses the challenges associated with web-scale pharmacovigilance.

Combining this with the widely used power of image analysis, Reece and Danforth mined through publicly accessible Instagram pictures trying to find predictive markers for depression [24]. They examined photos of 166 persons with a machine learning algorithm and found that certain features of the photos, such as the presence of faces, the use of filters, and the number of likes, could be used to predict depression. The results showed that the algorithm was able to accurately predict depression in 70% of the participants. The study concluded that Instagram photos can be used to predict depression and could be used to identify those at risk of developing the disorder. For example, if a person is posting photos that are dark, sad, or lonely, it could be an indication that they are struggling with depression. Additionally, if a person is posting fewer photos than usual, or if the photos they do post lack color or vibrancy, this could also be a sign of depression. Finally, if a person is posting photos of themselves that are heavily edited or filtered, this could be a sign that they are trying to hide their true emotions.

Artificial intelligence, clinical imaging, and multiomics

More examples of projects analyzing Big Data for medical purposes also rely on modern image analysis. Artificial intelligence (AI) and clinical imaging have a great potential to improve health care. AI/machine learning algorithms can be trained to detect specific markers in medical images, such as tumors, lesions, and other abnormalities. Big Data can provide the necessary data to train these algorithms, as well as to identify patterns and correlations in medical images. AI/machine learning algorithms can also be used to automate certain tasks, such as image segmentation and feature extraction, which can reduce the amount of time and effort required by clinicians. Additionally, AI/machine learning algorithms can be used to identify and diagnose diseases more accurately and quickly than traditional methods [25,26].

With the growing computational capabilities, Esteva et al. were successful in training a deep neural network model used to accurately classify skin cancer at a dermatologist-level accuracy [27]. The model was trained on a data set of 129,450 clinical images of 2032 different diseases, including skin cancer, and achieved an accuracy of 95.9% when tested on a separate data set of 600 clinical images of skin cancer. This is comparable to the accuracy of a board-certified dermatologist. The authors also discuss the potential implications of this model for the diagnosis and treatment of skin cancer. They suggest that the model could be used to reduce the time and cost associated with skin cancer diagnosis, as well as to improve the accuracy of diagnosis. Additionally, they suggest that the model could be used to identify patients who are at a higher risk of developing skin cancer and to provide personalized treatment plans.

Similar reliance on the power of image analysis is found in a project by Dhara et al. where they present a content-based image retrieval system for pulmonary nodules that assists radiologists in self-learning and diagnosis of lung cancer [28]. The system uses a combination of image processing techniques and machine learning algorithms to extract features from the images and classify them into benign and malignant nodules. The system is evaluated using a data set of CT scans from the Lung Image Database Consortium (LIDC), and the results show that the system is able to correctly classify the nodules with an accuracy of 97.5%. The paper also discusses the potential applications of the system in clinical practice and how it can be used to improve the accuracy of diagnosis of lung cancer.

While the last examples are concerned with generally improving health care through empowering diagnostics with computational guidance, more patient-centric personalized medicine and multiomics research has also seen a rise in publications and successful projects employing the power of Big Data-based analysis.

In the era of precision medicine, the integration of Big Data and multiomics approaches is becoming increasingly essential for gaining insights into the complex molecular mechanisms of various diseases. The integration of multiomics data facilitates the comprehensive characterization of the various molecular alterations in a given disease, which may aid in identifying novel drug targets or developing personalized treatments.

The combination of multiomics data with other clinical and phenotypic information enables researchers to develop more accurate and personalized diagnostic and treatment strategies. For instance, the glass patient concept involves the integration of various clinical and omics data for individual patients to develop personalized treatment strategies. This approach provides a more comprehensive understanding of the molecular alterations in an individual patient, enabling clinicians to make more informed decisions regarding diagnosis and treatment.

In conclusion, the integration of Big Data and multiomics approaches is revolutionizing health care by providing a more comprehensive understanding of the molecular mechanisms underlying various diseases. The integration of various omics data, clinical, and phenotypic information may aid in the development of personalized and effective diagnostic and treatment strategies [29,30].

Limitations of Big Data

While the use of Big Data and AI can provide significant advantages in the healthcare sector, it also presents potential drawbacks that must be considered. For instance, the storage and usage of extensive amounts of personal health data can expose individuals' confidential information to breaches [31−36]. Personal health information is sensitive and must be treated with utmost care to ensure patients' privacy rights are respected. Such breaches can lead to a loss of trust between healthcare providers and their patients, which can negatively affect patient engagement and willingness to participate in healthcare initiatives.

Moreover, the availability of such sensitive data can create blind spots in therapy. Data that is stored in large quantities can be challenging to interpret accurately, and the conclusions drawn from it can be misinterpreted, leading to errors in diagnosis or treatment [32,33,37,38]. As a result, healthcare providers must ensure that the data utilized in AI applications is appropriately screened, verified, and secured to provide reliable and valuable insights that can drive effective decision-making in clinical settings.

Additionally, the decision-making process of Big Data is often criticized due to the lack of transparency and accountability, as the presented solutions are generated by AI instead of human beings. The utilization of AI-generated solutions may lead to biased outcomes that can result in unequal access to healthcare services and inaccurate diagnoses and treatments. Moreover, it can lead to a lack of transparency in decision-making. The conclusions drawn from Big Data are often complex and difficult to understand, especially when decisions are made by machine learning models. This lack of transparency may lead to a lack of trust in the decision-making process, making patients reluctant to accept the decisions made by healthcare providers [32−34,37]. Additionally, the absence of accountability is a major concern in Big Data and AI decision-making processes, as solutions are produced by machines rather than human beings. Any errors or inaccuracies can only be attributed to the machine.

The quality of decisions that are generated by Big Data and AI depends entirely on the underlying data. If the data is biased, the resulting conclusions will be skewed, leading to biased outcomes based on certain criteria such as race, gender, and socioeconomic status [39]. As a result, access to healthcare services and treatments can become unequal, and diagnoses and treatments may be inaccurate. Therefore, healthcare providers must develop mechanisms that ensure that the decision-making process of Big Data is transparent and accountable. Strategies such as model explanations, algorithm auditing, and bias testing should be implemented to ensure that the outcomes of Big Data are unbiased, reliable, and transparent. Furthermore, the implementation of ethical guidelines for the use of AI in health care can ensure that AI-generated solutions adhere to ethical standards and respect patients' rights. Effective measures must be put in place to ensure that personal health information is safeguarded and that its usage is strictly monitored. Data protection protocols must be implemented, and security measures must be updated regularly to prevent security breaches. Additionally, transparent policies must be developed to ensure patients understand how their data is used and the specific benefits that they can derive from its utilization.

One notable effect of Big Data is the "Simpson Paradox," which is a phenomenon in probability and statistics. It occurs when a trend appears in several different groups of data but disappears or reverses when these groups are combined. For example, a study may show that a certain drug is effective in treating a certain condition in one group of patients, but when the same data is analyzed across all groups, the drug may appear to be ineffective. This can be caused by differences in the groups, such as age, gender, or other factors. By analyzing data from all sources, the true effectiveness of the drug can be determined, but neglecting the effectiveness for minor groups [40,41].

Another negative effect of Big Data is overfitting, which occurs when a model is too closely fit to the data, resulting in a model that is overly complex and not generalizable [31,32]. This can lead to predictions that are not reflective of the underlying data and can lead to incorrect conclusions. Overfitting can also lead to models that are too sensitive to small changes in the data, resulting in unreliable results.

The challenges associated with Big Data and its application to medicine are being studied, and strategies are being developed to address these issues. Adequate infrastructure for secure storage, data access, and sharing platforms, as well as data analysis and visualization platforms, are required. The development of standards and protocols for handling Big Data in collaboration with researchers, clinical professionals, and industry partners is essential for success. The National Cancer Institute's Genomic Data Commons, the National Institutes of Health's Cancer Genomics Cloud, and the European Bioinformatics Institute's Cancer Genomics Hub are already implementing these strategies and developing standards and protocols for data sharing and analysis [3,42,43].

However, the transformation of medicine requires more than just data alone. The integration of Big Data analysis techniques, including machine learning tools, neural networks, and deep learning, is crucial for the realization of personalized and precision medicine. Different approaches in machine learning can work synergistically to generate new predictions. Expert systems can integrate knowledge of general medical principles with data to apply this knowledge to new patients and data. Conversely, machines can learn rules from patient-level observations and identify combinations of variables that reliably predict outcomes. Machine learning will enhance different fields in medicine, such as radiology, immunohistochemistry, or anatomical pathology, through image analysis and pattern recognition. Moreover, machine learning can improve diagnosis accuracy by generating differential diagnosis and reducing overuse testing by suggesting high-value tests. The potential benefits of Big Data and machine learning to revolutionize health care are extensive, including the provision of more accurate and timely information to inform decisions, improve care, and reduce costs.

Conclusion

The healthcare industry has been increasingly leveraging the benefits of Big Data and AI in recent years. The ability to collect and analyze vast amounts of healthcare data has resulted in improved patient outcomes and cost savings. Big Data enables the development of predictive models that can identify patterns and make informed decisions, while AI techniques such as machine learning and deep learning provide powerful tools for data analysis, which can significantly improve clinical decision-making. These technologies have also resulted in the development of personalized/precision medicine approaches, which tailor treatment plans to individual patients, based on their unique health data. The use of Big Data and AI has also enabled the identification of new drug targets and the optimization of drug development processes.

To fully realize the potential of Big Data and machine learning in health care, it is important to continue to address the challenges and concerns surrounding their use and to work toward a collaborative approach between researchers, clinicians, and industry partners. By doing so, we can improve the accuracy, efficiency, and effectiveness of health care and ultimately improve patient outcomes.

References

[1] Cox M., Ellsworth D. Application-controlled demand paging for out-of-core visualization. In: Proceedings of the IEEE visualization conference 1997, 235−44. Available from: https://doi.org/10.1109/visual.1997.663888.

[2] Cirillo D, Valencia A. Big Data analytics for personalized medicine. Curr Opin Biotechnol 2019;58:161−7. Available from: https://doi.org/10.1016/J.COPBIO.2019. 03.004.

[3] Hulsen T, Jamuar SS, Moody AR, Karnes JH, Varga O, Hedensted S, et al. From Big Data to precision medicine. Front Med (Lausanne) 2019;6:34. Available from: https://doi.org/10.3389/FMED.2019.00034/BIBTEX.

[4] Mayer-Schönberger V, Ingelsson E. Big Data and medicine: a big deal? J Intern Med 2018;283:418−29. Available from: https://doi.org/10.1111/JOIM.12721.

[5] Li C, Chen Y, Shang Y. A review of industrial Big Data for decision making in intelligent manufacturing. Eng Sci Technol 2022;29:101021. Available from: https:// doi.org/10.1016/J.JESTCH.2021.06.001.

[6] Wieringa J, Kannan PK, Ma X, Reutterer T, Risselada H, Skiera B. Data analytics in a privacy-concerned world. J Bus Res 2021;122:915−25. Available from: https:// doi.org/10.1016/J.JBUSRES.2019.05.005.

[7] Quach S, Thaichon P, Martin KD, Weaven S, Palmatier RW. Digital technologies: tensions in privacy and data. J Acad Mark Sci 2022;50:1299−323. Available from: https://doi.org/10.1007/S11747-022-00845-Y/TABLES/4.

[8] Fröhlich H, Balling R, Beerenwinkel N, Kohlbacher O, Kumar S, Lengauer T, et al. From hype to reality: data science enabling personalized medicine. BMC Med 2018;16:1−15. Available from: https://doi.org/10.1186/S12916-018-1122-7/ FIGURES/5.

[9] Flores M, Glusman G, Brogaard K, Price ND, Hood L. P4 medicine: how systems medicine will transform the healthcare sector and society. J Per Med 2013;10(6):565. Available from: https://doi.org/10.2217/pme.13.57.

[10] Hood L, Balling R, Auffray C. Revolutionizing medicine in the 21st century through systems approaches. Biotechnology J 2012;7(8):992. Available from: https:// doi.org/10.1002/BIOT.201100306.

[11] Hood L, Heath JR, Phelps ME, Lin B. Systems biology and new technologies enable predictive and preventative medicine. Science (New York, N.Y.) 2004;306 (5696):640−3. Available from: https://doi.org/10.1126/SCIENCE.1104635.

[12] Maher B. Personal genomes: the case of the missing heritability. Nature 2008; 456:18−21. Available from: https://doi.org/10.1038/456018A.

[13] Manolio TA, Collins FS, Cox NJ, Goldstein DB, Hindorff LA, Hunter DJ, et al. Finding the missing heritability of complex diseases. Nature 2009;461:747−53. Available from: https://doi.org/10.1038/NATURE08494.

[14] Visscher PM, Wray NR, Zhang Q, Sklar P, McCarthy MI, Brown MA, et al. 10 Years of GWAS discovery: biology, function, and translation. Am J Hum Genet 2017;101:5−22. Available from: https://doi.org/10.1016/J.AJHG.2017.06.005.

[15] Kim H, Grueneberg A, Vazquez AI, Hsu S, de Los Campos G. Will Big Data close the missing heritability gap? Genetics 2017;207:1135−45. Available from: https:// doi.org/10.1534/GENETICS.117.300271/-/DC1.

[16] Yang C, Li C, Wang Q, Chung D, Zhao H. Implications of pleiotropy: challenges and opportunities for mining Big Data in biomedicine. Front Genet 2015;6:229. Available from: https://doi.org/10.3389/FGENE.2015.00229/BIBTEX.

[17] Young AI. Solving the missing heritability problem. PLoS Genet 2019;15:e1008222. Available from: https://doi.org/10.1371/JOURNAL.PGEN.1008222.

[18] Ollier W, Sprosen T, Peakman T. UK Biobank: from concept to reality. Pharmacogenomics 2005;6:639−46. Available from: https://doi.org/10.2217/ 14622416.6.6.639.

[19] Gulcher J, Stefansson K. Population genomics: laying the groundwork for genetic disease modeling and targeting. Clin Chem Lab Med 1998;36:523−7. Available from: https://doi.org/10.1515/CCLM.1998.089.

[20] Chen Z, Lee L, Chen J, Collins R, Wu F, Guo Y, et al. Cohort profile: the Kadoorie study of chronic disease in China (KSCDC. Int J Epidemiol 2005;34:1243−9. Available from: https://doi.org/10.1093/IJE/DYI174.

[21] Dugas AF, Hsieh YH, Levin SR, Pines JM, Mareiniss DP, Mohareb A, et al. Google flu trends: correlation with emergency department influenza rates and crowding metrics. Clin Infect Dis 2012;54:463−9. Available from: https://doi.org/10.1093/CID/CIR883.

[22] Ginsberg J, Mohebbi MH, Patel RS, Brammer L, Smolinski MS, Brilliant L. Detecting influenza epidemics using search engine query data. Nature 2009;457:1012−14. Available from: https://doi.org/10.1038/nature07634.

[23] White RW, Tatonetti NP, Shah NH, Altman RB, Horvitz E. Web-scale pharmacovigilance: listening to signals from the crowd. J Am Med Inf Assoc 2013;20:404. Available from: https://doi.org/10.1136/AMIAJNL-2012-001482.

[24] Reece AG, Danforth CM. Instagram photos reveal predictive markers of depression. EPJ Data Sci 2017;6:15. Available from: https://doi.org/10.1140/epjds/s13688-017-0110-z.

[25] Ahmad Z, Rahim S, Zubair M, Abdul-Ghafar J. Artificial intelligence (AI) in medicine, current applications and future role with special emphasis on its potential and promise in pathology: present and future impact, obstacles including costs and acceptance among pathologists, practical and philosophical considerations. A Compr Rev Diagn Pathol 2021;16. Available from: https://doi.org/10.1186/S13000-021-01085-4.

[26] Athanasopoulou K, Daneva GN, Adamopoulos PG, Scorilas A. Artificial intelligence: the milestone in modern biomedical research. BioMedInformatics 2022;2:727−44. Available from: https://doi.org/10.3390/BIOMEDINFORMATICS2040049 2022, 2, 727−744.

[27] Esteva A, Kuprel B, Novoa RA, Ko J, Swetter SM, Blau HM, et al. Dermatologist-level classification of skin cancer with deep neural networks. Nature 2017;542:115−18. Available from: https://doi.org/10.1038/NATURE21056.

[28] Dhara AK, Mukhopadhyay S, Dutta A, Garg M, Khandelwal N. Content-based image retrieval system for pulmonary nodules: assisting radiologists in self-learning and diagnosis of lung cancer. J Digit Imaging 2017;30:63−77. Available from: https://doi.org/10.1007/S10278-016-9904-Y.

[29] Hasin Y, Seldin M, Lusis A. Multi-omics approaches to disease. Genome Biol 2017;18(1):1−15. Available from: https://doi.org/10.1186/S13059-017-1215-1.

[30] Prosperi M, Min JS, Bian J, Modave F. Big Data hurdles in precision medicine and precision public health. BMC Med Inf Decis Mak 2018;18:1−15. Available from: https://doi.org/10.1186/S12911-018-0719-2/FIGURES/7.

[31] Obermeyer Z, Emanuel EJ. Predicting the future—Big Data, machine learning, and clinical medicine. N Engl J Med 2016;375:1216. Available from: https://doi.org/10.1056/NEJMP1606181.

[32] Alyass A, Turcotte M, Meyre D. From Big Data analysis to personalized medicine for all: challenges and opportunities. BMC Med Genomics 2015;8:1−12. Available from: https://doi.org/10.1186/S12920-015-0108-Y/FIGURES/6.

[33] Knoppers BM, Thorogood AM. Ethics and Big Data in health. Curr Opin Syst Biol 2017;4:53−7. Available from: https://doi.org/10.1016/J.COISB.2017.07.001.

[34] Kraus JM, Lausser L, Kuhn P, Jobst F, Bock M, Halanke C, et al. Big Data and precision medicine: challenges and strategies with healthcare data. Int J Data Sci Analyt 2018;6:241−9. Available from: https://doi.org/10.1007/S41060-018-0095-0.

[35] Galetsi P, Katsaliaki K, Kumar S. Big Data analytics in health sector: theoretical framework, techniques and prospects. Int J Inf Manage 2020;50:206−16. Available from: https://doi.org/10.1016/J.IJINFOMGT.2019.05.003.

[36] Frakt A.B., Pizer S.D. The promise and perils of Big Data in healthcare; 2016. https://doi.org/10.1056/NEJMp1108726.

[37] Beckmann JS, Lew D. Reconciling evidence-based medicine and precision medicine in the era of Big Data: challenges and opportunities. Genome Med 2016;8:1—11. Available from: https://doi.org/10.1186/S13073-016-0388-7/TABLES/1.

[38] Shafqat S, Kishwer S, Ur Rasool R, Qadir J, Amjad T, Farooq Ahmad H. Big Data analytics enhanced healthcare systems: a review. J Supercomput 2020;76:1754—99. Available from: https://doi.org/10.1007/s11227-017-2222-4.

[39] Norori N, Hu Q, Aellen FM, Faraci FD, Tzovara A. Addressing bias in Big Data and AI for health care: a call for open science. Patterns 2021;2:100347. Available from: https://doi.org/10.1016/J.PATTER.2021.100347.

[40] Wagner CH. Simpson's paradox in real life. Am Stat 1982;36:46. Available from: https://doi.org/10.2307/2684093.

[41] Pearl J. Understanding Simpson's paradox. SSRN Electron J 2013;. Available from: https://doi.org/10.2139/SSRN.2343788.

[42] Batko K, Ślęzak A. The use of Big Data analytics in healthcare. J Big Data 2022;9. Available from: https://doi.org/10.1186/S40537-021-00553-4.

[43] Dash S, Shakyawar SK, Sharma M, Kaushik S. Big Data in healthcare: management, analysis and future prospects. J Big Data 2019;6:1—25. Available from: https://doi.org/10.1186/S40537-019-0217-0/FIGURES/6.

CHAPTER 6

Artificial intelligence and personalized medicine: transforming patient care

Marc Ghanem[1], Abdul Karim Ghaith[2,3] and Mohamad Bydon[2,3]

[1]Gilbert and Rose-Marie Chagoury School of Medicine, Lebanese American University, Beirut, Lebanon
[2]Mayo Clinic Neuro-Informatics Laboratory, Mayo Clinic, Rochester, MN, United States
[3]Department of Neurologic Surgery, Mayo Clinic, Rochester, MN, United States

Introduction

Artificial intelligence (AI) has emerged as a powerful force in various industries, including health care, paving the way for more personalized and precise medicine [1]. This chapter will delve into the transformative role of AI in personalized medicine and its immense potential to revolutionize the future of health care. By leveraging cutting-edge AI technologies, personalized medicine aims to provide tailored healthcare interventions that consider individual patients' unique genetic makeup, lifestyle, and environmental factors, ultimately improving patient outcomes and overall healthcare quality [2]. The role of AI in personalized medicine encompasses a wide array of applications, from more accurate diagnostics and risk assessments to personalized treatment plans and disease prediction [3]. Employing machine learning (ML), deep learning (DL), and natural language processing (NLP), AI has the power to analyze vast amounts of complex data and unlock novel insights that can directly impact patient care [4]. By augmenting the capabilities of healthcare professionals and enabling more informed decision-making, AI is set to transform the way medicine is practiced, paving the way for a more patient-centric, efficient, and precise healthcare system [4]. This chapter delves into the various aspects of AI-driven personalized medicine, highlighting its key applications, ongoing challenges, and promising research efforts. Join us as we embark on this journey to explore the transformative potential of AI in shaping the future of personalized medicine and health care.

The New Era of Precision Medicine
DOI: https://doi.org/10.1016/B978-0-443-13963-5.00012-1
© 2024 Elsevier Inc.
All rights reserved.

Foundations of artificial intelligence in personalized medicine

Brief overview of artificial intelligence technologies and techniques

Machine learning

ML pertains to a particular domain of AI, emphasizing creating algorithms capable of acquiring knowledge and formulating predictions based on data. The techniques utilized in ML encompass training models to detect patterns and associations within data sets, thus enabling them to make informed decisions or projections without explicit programming. Utilizing ML algorithms for personalized medicine can enhance the comprehension of patient information, pinpoint trends, and produce significant revelations that will aid in medical assessments, predictions, and decision-making. Gligorijevic et al. discussed various methods for integrating biological data from different sources, including ML techniques. The authors argue that ML can be particularly useful for integrating large and complex data sets, and provide several examples of how ML has been used to analyze biological data [5]. Another study by Ho et al. highlights the recent ML application developments and described how ML approaches could lead to improved complex disease prediction, which will help to incorporate genetic features into future personalized health care. The authors also discussed the future application of ML prediction models that might help manage complex diseases by providing tissue-specific targets for customized, preventive interventions [6].

MacEachern et al. published a study investigating ML utilization for precision medicine's "big data," in the context of genetics and genomics. They concluded that both fields have been essential to the evolution of precision medicine and are well suited to ML technology [7].

Deep learning

DL is a distinct variant of ML that utilizes artificial neural networks (ANNs) to imitate intricate patterns and representations identified in data. ANNs emulate the design and functionality of our brain by integrating interconnected layers filled with nodes or neurons for receiving and transmitting the information [8]. Proficient in processing vast data sets containing various dimensions, DL models are particularly advantageous for personalized medicine applications such as decoding genomic data and analyzing medical imaging outcomes. Martorell-Marugán et al. provided an overview of the main applications of DL methods in biomedical research, focusing on omics data analysis and precision medicine applications. They also discussed the implementations for improving the diagnosis, treatment, and classification of complex diseases [9]. Another study by Grapov et al. examined the challenges and opportunities for DL at the systems and biological scale for a precision medicine readership. The authors concluded that the rise of DL in genomic, proteomic, and metabolomic data integration has the potential to transform precision medicine. By analyzing large and complex data sets using DL algorithms, researchers can identify biomarkers and develop personalized treatment plans based on an individual's unique genomic and molecular profile. This approach has the potential to significantly improve patient outcomes and reduce healthcare costs by providing targeted and effective treatments. However, there are still challenges to overcome, including data quality and privacy concerns, as well as the need for standardized methods and data sharing. Future research in this area should focus on developing robust and validated DL models that can be integrated into clinical practice to improve patient care [10].

Natural language processing
NLP is a form of AI that enables computers to understand, interpret, and even generate human language [11]. Using NLP technology, AI systems can derive valuable insights from unstructured textual data in electronic health records (EHRs) or medical literature. This technique has proven particularly useful in personalized medicine as it identifies important patient traits and extracts relevant medical information while promoting better communication between healthcare providers and their patients.

Artificial intelligence-driven diagnostic tools and applications
Medical imaging analysis
AI showcases exceptional capabilities in the domain of medical imaging analysis, leading to more precise and rapid interpretation of images, such as X-rays, CT scans, and MRIs [12]. AI algorithms can be trained to recognize distinct patterns, irregularities, and structures within images, aiding healthcare professionals in diagnosing various conditions. AI-assisted medical imaging analysis expedites the diagnostic process and minimizes the possibility of human error, ultimately enhancing patient outcomes. Gore published the historical evolution of some major changes in radiology that are traced as background to how AI may also be embraced into practice.

Potential new capabilities provided by AI offer exciting prospects for more efficient and effective use of medical images [13].

Genomic data interpretation

Genomic data has surged due to the development of next-generation sequencing technology, providing important new insights into the genetic causes of disease and individual responses to therapy [14]. AI critically analyzes this vast and complex data and identifies disease-related genetic variants, possible therapeutic targets, and patient-specific therapy recommendations. AI-guided approaches can offer a more comprehensive picture of a patient's health and encourage the development of completely tailored treatment regimens by combining genetic data with other patient information [15].

Integrating different data modalities provides opportunities to increase the robustness and accuracy of diagnostic and prognostic models, bringing AI closer to clinical practice. Lipkova et al. supported these advances by presenting a synopsis of AI methods and strategies for multimodal data fusion and association discovery. They concluded that AI has the potential to revolutionize cancer diagnosis and treatment by combining data from different sources, such as genomics, radiomics, and EHRs. This approach can facilitate the early detection of cancer, prediction of treatment response, and identification of new therapeutic targets. Biomarker discovery has been enhanced by the high-throughput technologies of omics [16].

Biomarker identification

Biomarkers are quantifiable biological indicators that convey important details about a person's health, the course of the disease, or the response to treatment [17]. Large databases may be mined by AI algorithms for possible biomarkers, assisting in discovering new therapeutic, prognostic, and diagnostic targets. By uncovering previously unrecognized biomarkers, AI can assist in implementing more targeted therapies and enabling health professionals to make more informed decisions about care delivery, ultimately leading to better patient outcomes. A study by Michelhaugh et al. studied the advancements in technology that improved the biomarker discovery in the field of heart failure (HF). The study showed that using omics and AI in biomarker discovery can aid clinicians by identifying markers of risk for developing HF, monitoring care, determining prognosis, and developing druggable targets [15].

Artificial intelligence for personalized drug therapy

Pharmacogenomics

Pharmacogenomics is integral to personalized medicine, as it explores how a person's genetic composition impacts their drug response [15]. AI can contribute significantly by analyzing genomic data and detecting genetic variances that might influence drug response. The obtained data may then be used to tailor more effective pharmacological regimens [18]. Jeibouei et al. report and discuss the molecular classification of breast cancer subtypes and the current drug regimens with consideration of pharmacogenomics in response and resistance to treatment [19].

Precision dosing

AI may also contribute to precision dosing by individually determining the optimal drug dose for patients. AI algorithms can determine the most effective dose for a patient by examining patient-specific characteristics such as age, weight, genetics, and organ function.

By reducing patient side effects and ameliorating treatment outcomes, a personalized drug therapy approach to health care advances patient outcomes and enhances healthcare systems' productivity and efficiency [20]. A study published by Tyson et al. revealed the importance of considering more prompt and precise dosing delivery to high-priority patients. They discussed variables to consider when prioritizing precision dosing candidates while highlighting the key examples of precision dosing that have been successfully used to improve patient care [21].

Artificial intelligence in disease prediction and prevention

Artificial intelligence-powered risk assessment

Predictive models for disease susceptibility

AI models can predict individual patient disease susceptibilities by assessing large and diverse data sets, including genetic, environmental, and lifestyle factors [22]. These models can help identify people at a higher risk of contracting specific diseases, allowing for early intervention and targeted prevention measures. AI-powered risk assessments would allow healthcare practitioners to provide more preventive and customized care, reducing illness burden and ameliorating population health [23]. Collins et al. discuss the application of genomic and proteomic technologies in predictive, preventive, and personalized medicine. They use type 1 diabetes (T1D) as an example to discuss pertinent issues related to high-throughput bio-marker discovery [24].

Early detection of disease patterns

AI can also diagnose diseases early on by detecting subtle changes in patient data, including medical imaging, EHR, and biomarker levels [23]. Early detection can result in more effective management, enhancing patient outcomes and reducing the expenses associated with advanced ill-ness management. AI can shift patient management from reactive therapy to proactive prevention by aiding in the early detection of disease trends [25]. Mathema et al. focus on the recent development of ML/DL meth-ods to provide integrative solutions in discovering cancer-related biomar-kers, and their utilization in precision medicine [26].

Artificial intelligence-driven lifestyle recommendations

Personalized nutrition and exercise plans

AI can assist in developing individualized nutrition and exercise programs by assessing individual patient data such as genetic information, metabolic profiles, past injuries, and lifestyle habits [27]. These personalized recom-mendations can assist individuals in optimizing their health and lowering their chance of developing chronic conditions such as obesity, diabetes, or cardiovascular disease. AI-powered lifestyle suggestions can enable indivi-duals to take charge of their health and well-being by providing tailored nutrition and physical workout guidelines [28]. Mortazavi et al. review new programs that can generate personalized/precision nutrition recom-mendations based on measurements of gut microbiota and continuous glucose monitors with AI [29].

Behavioral health monitoring and support

AI can help monitor and support behavioral health by analyzing a patient's self-reported assessments, social media activity, or smartphone usage habits [30]. By detecting early symptoms of mental health issues or risky conduct, AI systems can trigger interventions such as giving support services or alerting healthcare providers. This preventive strategy can help prevent mental health disorders from worsening and promote well-being [31]. Pham et al. explore the potential impact of AI chatbots on psychia-try, highlighting emerging interventions such as conversational therapy bots. These interventions aim to teach emotional coping mechanisms, provide support for those with communication difficulties, and advance the field of digital psychiatry [32].

Wearable devices and remote patient monitoring

Nowadays, wearable devices and remote patient monitoring systems continuously gather patient health data that AI models can interpret afterward [33]. The integrated models can detect anomalies by analyzing health parameters such as heart rate, blood pressure, and glucose levels. These wearables offer patients' real-time feedback to help individuals better understand their health. Such technologies can ultimately contribute to disease prevention and improved health outcomes [34]. Sujith et al. present a systematic review of the use of smart health monitoring (SHM) systems to address the challenges of disease prevention and control, particularly in light of the technological advancements of industry 5.0 and 5G. They discuss the benefits of cost-effective sensors, blockchain integration for improved data security, and the use of DL and ML for early disease detection and preventive health care [35].

Artificial intelligence and patient-centered care

Personalized patient engagement

Artificial intelligence-powered chatbots and virtual assistants

AI-powered chatbots and virtual assistants have the potential to improve patient engagement by offering personalized support and guidance through digital platforms [36]. These tools can assist patients in various ways, such as answering their questions, reminding them of appointments, and providing personalized health advice. Ashwini et al. explore the use of telemedicine to provide healthcare services during the COVID-19 pandemic. The authors develop a NLP-based multilingual conversational bot that provides chronic patients with fundamental healthcare education and information [37].

Tailored patient education materials

AI can generate individualized patient education materials based on a person's unique needs, preferences, and learning styles [38]. This process would have AI algorithms find the most relevant and understandable educational information based on the patient's request. This targeted approach can offer patients the knowledge and resources they need to understand their diseases better and make informed healthcare decisions.

Enhancing patient–provider communication

Artificial intelligence-enabled decision support systems for clinicians

AI-driven decision support systems can push for more patient-centered treatment [39]. These systems can examine medical literature, clinical guidelines, and patient data to provide recommendations specific to each patient's needs [40]. These technologies pave the way for more individualized treatment decisions by providing physicians with immediate access to patient-specific data and evidence-based suggestions. Loftus et al. discuss the potential benefits of AI-enabled decision support systems in nephrology, including their ability to predict acute kidney injury, identify risk factors for chronic kidney disease, and aid in decision-making following renal transplantation [41].

Ethical and legal considerations in artificial intelligence and personalized medicine

Ensuring data privacy and security

Data privacy and security are crucial when incorporating AI into personalized medicine [42]. The increasing collection and analysis of sensitive patient information necessitate strong measures to protect patient privacy and ensure health data confidentiality. Healthcare providers, technology developers, and regulators must collaborate to establish and adhere to strict data protection policies, encryption protocols, and access controls to safeguard patient information and prevent unauthorized use or disclosure [43].

Addressing algorithmic bias and fairness

AI algorithms may unintentionally perpetuate biases or inequities in health care if trained on biased or unrepresentative data [44]. The development and validation of AI models must consider potential sources of bias to ensure justice and equity in AI-driven individualized medicine. This includes using diversified and representative data and regularly reviewing and optimizing AI platforms to guarantee accurate predictions and recommendations for all patient populations [45].

Regulatory landscape and guidelines

A thorough regulatory framework that guarantees the safety, effectiveness, and moral application of AI technologies in health care must be established and maintained as AI continues to make substantial advancements in personalized medicine [46]. This entails creating precise norms and standards for AI development, validation, deployment, and strong systems for ongoing performance monitoring and postmarket oversight. Specific guidelines for healthcare AI are already being developed by regulatory organizations like the FDA in the United States [47]. This ongoing process will require a fostering collaboration between AI researchers, healthcare professionals, and policymakers to navigate the complexities of AI integration in personalized medicine.

Future directions and challenges

Artificial intelligence-driven innovations in personalized medicine

As AI continues to evolve and mature, it is anticipated that innovations will emerge in personalized medicine, further enhancing diagnostic accuracy, treatment effectiveness, and patient-centered care [48]. AI-driven innovations may include the development of more sophisticated algorithms, the integration of multimodal data sources, and the exploration of novel therapeutic targets. By embracing these advances, personalized medicine has the potential to revolutionize health care and improve patient outcomes [49].

Overcoming technical, clinical, and societal barriers

It is anticipated that new developments in personalized medicine will appear as AI continues to develop and mature, significantly improving diagnostic precision, therapeutic efficacy, and patient-centered care. With the upcoming developments of increasingly complex algorithms and the incorporation of multimodal data sources, AI will likely transform health care and enhance individualized patient management soon.

Conclusion

In this chapter, we have explored the various ways AI is changing personalized medicine, including its applications in diagnostics, drug discovery, disease prediction, prevention, and patient-centered care. We have also discussed the ethical and legal considerations surrounding the use of AI in health care, as well as the future directions and challenges that lie ahead. These key points highlight the increasingly significant role AI plays in transforming personalized medicine and shaping the future of health care.

References

[1] Jiang F, Jiang Y, Zhi H, et al. Artificial intelligence in healthcare: past, present and future. Stroke Vasc Neurol 2017;2(4):230−43.

[2] Bodiroga-Vukobrat N, Rukavina D, Pavelić K, Sander GG. Personalized medicine in healthcare systems: legal. Medical and economic implications. Springer; 2019.

[3] Mahler M. Precision medicine and artificial intelligence: the perfect fit for autoimmunity. Academic Press; 2021.

[4] Ahuja AS. The impact of artificial intelligence in medicine on the future role of the physician. PeerJ. 2019;7:e7702.

[5] Gligorijević V, Pržulj N. Methods for biological data integration: perspectives and challenges. J R Soc Interface 2015;12(112). Available from: https://doi.org/10.1098/rsif.2015.0571.

[6] Ho DSW, Schierding W, Wake M, Saffery R, O'Sullivan J. Machine learning SNP based prediction for precision medicine. Front Genet 2019;10:267.

[7] MacEachern SJ, Forkert ND. Machine learning for precision medicine. Genome. 2021;64(4):416−25.

[8] Alzubaidi L, Zhang J, Humaidi AJ, et al. Review of deep learning: concepts, CNN architectures, challenges, applications, future directions. J Big Data 2021;8(1):53.

[9] Martorell-Marugán J, Tabik S, Benhammou Y, et al. Deep learning in omics data analysis and precision medicine. In: Husi H, editor. Computational biology. Codon Publications; 2019.

[10] Grapov D, Fahrmann J, Wanichthanarak K, Khoomrung S. Rise of deep learning for genomic, proteomic, and metabolomic data integration in precision medicine. OMICS 2018;22(10):630−6.

[11] Luo JW, Chong JJR. Review of natural language processing in radiology. Neuroimaging Clin N Am 2020;30(4):447−58.

[12] Rubin DL. Artificial intelligence in radiology, an issue of radiologic clinics of North America, 59. Elsevier; 2021.

[13] Gore JC. Artificial intelligence in medical imaging. Magn Reson Imaging 2020;68:A1−4.

[14] Beigh MM. Next-generation sequencing: the translational medicine approach from "bench to bedside to population". Medicines (Basel) 2016;3(2). Available from: https://doi.org/10.3390/medicines3020014.

[15] Michelhaugh SA, Januzzi Jr. JL. Using artificial intelligence to better predict and develop biomarkers. Clin Lab Med 2023;43(1):99−114.

[16] Lipkova J, Chen RJ, Chen B, et al. Artificial intelligence for multimodal data integration in oncology. Cancer Cell 2022;40(10):1095−110.

[17] Seyhan AA, Carini C. Are innovation and new technologies in precision medicine paving a new era in patients centric care? J Transl Med 2019;17(1):114.

[18] Abdelhalim H, Berber A, Lodi M, et al. Artificial intelligence, healthcare, clinical genomics, and pharmacogenomics approaches in precision medicine. Front Genet 2022;13:929736.

[19] Jeibouei S, Akbari ME, Kalbasi A, et al. Personalized medicine in breast cancer: pharmacogenomics approaches. Pharmgenomics Pers Med 2019;12:59−73.

[20] Paterick TE, Patel N, Tajik AJ, Chandrasekaran K. Improving health outcomes through patient education and partnerships with patients. Proceedings 2017;30(1):112−13.

[21] Tyson RJ, Park CC, Powell JR, et al. Precision dosing priority criteria: drug, disease, and patient population variables. Front Pharmacol 2020;11:420.

[22] Ahmed Z, Mohamed K, Zeeshan S, Dong X. Artificial intelligence with multifunctional machine learning platform development for better healthcare and precision medicine. Database 2020;2020. Available from: https://doi.org/10.1093/database/baaa010.

[23] Samal L, Fu HN, Camara DS, Wang J, Bierman AS, Dorr DA. Health information technology to improve care for people with multiple chronic conditions. Health Serv Res 2021;56(Suppl. 1):1006−36.

[24] Collins CD, Purohit S, Podolsky RH, et al. The application of genomic and proteomic technologies in predictive, preventive and personalized medicine. Vasc Pharmacol 2006;45(5):258−67.

[25] Uzun Ozsahin D, Ikechukwu Emegano D, Uzun B, Ozsahin I. The systematic review of artificial intelligence applications in breast cancer diagnosis. Diagnostics (Basel) 2022;13(1). Available from: https://doi.org/10.3390/diagnostics13010045.

[26] Mathema VB, Sen P, Lamichhane S, Orešič M, Khoomrung S. Deep learning facilitates multi-data type analysis and predictive biomarker discovery in cancer precision medicine. Comput Struct Biotechnol J 2023;21:1372−82.

[27] Subramanian M, Wojtusciszyn A, Favre L, et al. Precision medicine in the era of artificial intelligence: implications in chronic disease management. J Transl Med 2020;18 (1):472.

[28] Lazarus S. Artificial intelligence could help work out the best diet for every individual. CNN, https://www.cnn.com/2022/12/20/world/artificial-intelligence-nutrition-hnk-spc-intl/index.html; Published December 21, 2022 [accessed 01.04.23].

[29] Mortazavi BJ, Gutierrez-Osuna R. A review of digital innovations for diet monitoring and precision nutrition. J Diabetes Sci Technol 2023;17(1):217−23.

[30] Chen IM, Chen YY, Liao SC, Lin YH. Development of digital biomarkers of mental illness via mobile apps for personalized treatment and diagnosis. J Pers Med 2022;12(6). Available from: https://doi.org/10.3390/jpm12060936.

[31] Min JA, Lee CU, Lee C. Mental health promotion and illness prevention: a challenge for psychiatrists. Psychiatry Investig 2013;10(4):307−16.

[32] Pham KT, Nabizadeh A, Selek S. Artificial intelligence and chatbots in psychiatry. Psychiatr Q 2022;93(1):249−53.

[33] Sabry F, Eltaras T, Labda W, Alzoubi K, Malluhi Q. Machine learning for healthcare wearable devices: the big picture. J Healthc Eng 2022;2022:4653923.

[34] Lu L, Zhang J, Xie Y, et al. Wearable health devices in health care: narrative systematic review. JMIR Mhealth Uhealth 2020;8(11):e18907.

[35] Sujith AVLN, Sajja GS, Mahalakshmi V, et al. Review of smart health monitoring using deep learning and artificial intelligence. Neurosci Inform 2022;2(3):100028.

[36] Jadczyk T, Wojakowski W, Tendera M, Henry TD, Egnaczyk G, Shreenivas S. Artificial intelligence can improve patient management at the time of a pandemic: the role of voice technology. J Med Internet Res 2021;23(5):e22959.

[37] AshwiniS., Rajalakshmi N.R., Victor Paul P., Jayakumar L. Dynamic NLP enabled chatbot for rural health care in India. In: 2022 second international conference on computer science, engineering and applications (ICCSEA); 2022. p. 1−6.

[38] Masters K. Ethical use of artificial intelligence in health professions education: AMEE Guide No.158. Med Teach, 13. 2023. p. 1−11.

[39] Juang WC, Hsu MH, Cai ZX, Chen CM. Developing an AI-assisted clinical decision support system to enhance in-patient holistic health care. PLoS One 2022;17 (10):e0276501.

[40] Sim I, Gorman P, Greenes RA, et al. Clinical decision support systems for the practice of evidence-based medicine. J Am Med Inf Assoc 2001;8(6):527−34.

[41] Loftus TJ, Shickel B, Ozrazgat-Baslanti T, et al. Artificial intelligence-enabled decision support in nephrology. Nat Rev Nephrol 2022;18(7):452−65.

[42] Forcier MB, Gallois H, Mullan S, Joly Y. Integrating artificial intelligence into health care through data access: can the GDPR act as a beacon for policymakers? J Law Biosci 2019;6(1):317−35.

[43] McGraw D, Mandl KD. Privacy protections to encourage use of health-relevant digital data in a learning health system. NPJ Digit Med 2021;4(1):2.

[44] Gianfrancesco MA, Tamang S, Yazdany J, Schmajuk G. Potential biases in machine learning algorithms using electronic health record data. JAMA Intern Med 2018;178 (11):1544−7.

[45] Klumpp M, Hintze M, Immonen M, et al. Artificial intelligence for hospital health care: application cases and answers to challenges in European Hospitals. Healthcare (Basel) 2021;9(8). Available from: https://doi.org/10.3390/healthcare9080961.

[46] Bélisle-Pipon JC, Couture V, Roy MC, Ganache I, Goetghebeur M, Cohen IG. What makes artificial intelligence exceptional in health technology assessment? Front Artif Intell 2021;4:736697.

[47] Vokinger KN, Gasser U. Regulating AI in medicine in the United States and Europe. Nat Mach Intell 2021;3(9):738−9.

[48] Subbiah V. The next generation of evidence-based medicine. Nat Med 2023;29 (1):49−58.

[49] Goetz LH, Schork NJ. Personalized medicine: motivation, challenges, and progress. Fertil Steril 2018;109(6):952−63.

CHAPTER 7

Ethical considerations in precision medicine

Lisa S. Parker[1,2]
[1]Center for Bioethics & Health Law, University of Pittsburgh, Pittsburgh, PA, United States
[2]Department of Human Genetics, University of Pittsburgh, Pittsburgh, PA, United States

Introduction

Precision medicine (PM) is characterized as "delivering the right treatments, at the right time, every time to the right person" [1]. That its name could mislead and create false expectations among patients and the public is perhaps the first ethical worry it presents. It is important to have an accurate understanding of how the knowledge base for PM is developed in order to have realistic expectations of what PM may offer individual patients in particular contexts.

Despite the term "precision," or "personalized medicine," as it was previously phrased, PM is epidemiologically driven: "The science that associates particular molecular markers with different risks, outcomes, and clinical indications is population-based, to begin with" [2, p. 24]. It "does not literally mean the creation of drugs or medical devices that are unique to a patient, but rather the ability to classify individuals into subpopulations that differ in their susceptibility to a particular disease or their response to a specific treatment" and to concentrate "preventive and therapeutic interventions ... on those who will benefit, sparing expense and side effects for those who will not" [3, p. 7]. The care of individual patients is based on what was found previously to be true of—and beneficial for—a population of relevantly similar patients. In this way, PM is a more refined or precise stage of contemporary evidence-based medicine. Interventions and clinical guidelines based on studies constituting the knowledge base for PM still hold only a population-based probability of benefitting an individual patient who is relevantly similar to the studied population. Individual patients' experience with the resulting recommended intervention—drug, preventive or treatment regimen, or behavior change—will still vary.

PM capitalizes on the evolving understanding of human traits and conditions resulting from the study of genetics, genomics, and other omics

The New Era of Precision Medicine
DOI: https://doi.org/10.1016/B978-0-443-13963-5.00002-9
© 2024 Elsevier Inc.
All rights reserved.
143

and employs technologies emerging from those fields (e.g., exome and genome sequencing, or gene editing). It incorporates, yet is broader than, genomic medicine. Currently in PM, when classifying people into populations that differ in disease susceptibility or likely treatment response, most attention seems to focus on genetic differences. Because most traits and conditions result from the interaction of genes with environments, as it matures, PM should increasingly take into account environmental, behavioral, and lifestyle factors, as well as genotypes.

This chapter focuses on ethical issues that arise in PM because of its use of genetic testing, genomic sequencing, and molecular profiling, though it also glances ahead to a time when environmental factors are similarly used to categorize patients, predict the risk of disease and treatment response, and develop targeted interventions. The ethical issues are discussed in relation to the domains of current PM practice in which they arise most obviously, though most are relevant to all arenas of PM.

The chapter begins with the diagnosis and treatment of rare diseases to discuss ethical and policy considerations presented by very expensive treatments and the risk that PM may actually worsen the problem of "orphan diseases." Second, it considers issues with regard to precision oncology. These issues include disparities in health and health care, informed consent for clinical care and research, management of incidental findings, the rights to-know and not-to-know health-related information about oneself, whether there is a duty to warn or protect nonpatient third parties, and intrafamilial disclosure of hereditary risks. The third section discusses ethical considerations arising most prominently with regard to polygenic risk assessment: risks of discrimination, stigmatization, and blame, as well as issues of justice arising because people of European ancestry constitute the vast majority of those studied in genetic/genomic studies that form the foundation for PM. The fourth section focuses on pharmacogenomics (PGx) and examines its benefits for drug development and its potential liabilities and benefits for patients and their access to treatments.

The diagnostic odyssey and quest for treatment of rare conditions: economics and policy lessons for precision medicine

Patients, families, healthcare practitioners, and politicians face tremendous obstacles in the diagnosis and treatment of rare illnesses. Those with rare conditions often endure a long period of visiting different specialists and having multiple rounds of testing before they receive a diagnosis or information about the cause of their symptoms. Ending this diagnostic odyssey is often considered valuable in itself, as it relieves anxiety associated with uncertainty. (A diagnosis can also inform reproductive decisions for patients or their families, which may relieve or create anxiety, depending on their values and life goals.) Ending the diagnostic odyssey, however, often marks the beginning of a quest for effective treatment [4]. Identifying the genetic contributions to various rare (and indeed, many common) conditions has not led to the development of treatments as directly as was hoped [5–13].

Attempts to address the needs of those with rare conditions predate the era of PM. The US Orphan Drug Act of 1983 was passed to incentivize the development of treatments for these "orphan conditions" for which market-based incentives are limited by the small market size of those who would benefit from a company's investment [14]. The Act defines a rare condition as one that affects fewer than 200,000 people in the United States. In addition to a lack of market incentives, having only small pools of potential study participants also limits research: For many rare cancers, there is an insufficient number of study-eligible patients to enroll in existing studies [15]. The rapid course of some rare conditions may also limit research on them: For example, 30% of children with a rare condition do not live to the age of 6 [16].

Challenges to enrolling sufficient participants to conduct treatment trials for rare conditions provide lessons for PM, as the condition + genotype subgroups that it creates may face similar challenges with regard to research design and limited incentives to pursue research. PM may actually create new healthcare disparities by identifying rare genetic etiologies for some patients' conditions: "For example, someone with lung cancer may be seen as ineligible for some kinds of treatment or research because she is part of the roughly 5% of nonsmall cell lung cancers with an ALK gene rearrangement" [4, p. 837]. PM may create "orphan genotypes," condition + genotypes for which there is no effective treatment and little market-based incentive to develop one.

Moreover, effective treatments that are developed for rare conditions have often proved prohibitively expensive, leading to headlines like "Nations Face Crushing Cost of Medical Miracles" [17]. Some cost hundreds of thousands of dollars each year for a patient's lifetime—e.g., nusinersen for treatment of a rare recessive neuromuscular condition, spinal muscular atrophy, at $370,000/year after the first year at $750,000 [4], or enzyme replacement therapy, peaking at $450,000, for mucopolysaccharidosis VI, a very rare lysosomal storage disorder [18]. The gene-therapy Zolgensma costs $1.7 million for a one-time infusion to treat spinal muscular atrophy [17].

When PM research develops treatment for subpopulations of patients characterized by a specific condition + genotype, policymakers, the Centers for Medicare and Medicaid, and health insurers will need to determine how to fund their high cost. They are likely to be orders of magnitude more expensive than previous-generation interventions. In the absence of government or health insurance coverage, such high costs limit access to either those with substantial wealth or those who qualify for manufacturer-provided treatment. Families are forced to make difficult decisions about whether to incur astronomical costs of treatment, including weighing the competing interests of their children when one of them has a rare condition. The path to diagnosis and treatment may be exorbitant and emotionally draining, with patients and their families frequently having to navigate complex healthcare systems and insurance policies.

To summarize, PM provides hope for individuals suffering from rare diseases, but its development and implementation present economic and policy problems that must be addressed. To ensure availability and affordability for all patients, policymakers might consider a variety of initiatives, including giving incentives for medication research, supporting market competition, and investing in diagnostic tools and infrastructure.

Precision oncology: considerations of informed consent, incidental findings, and the "Duty to Warn"

Precision oncology is a method of treating cancer that involves tailoring a patient's course of treatment based on genetic and molecular data. In sharp contrast to rare conditions, cancer is the second-most common disease in the United States, when all types of cancer are considered together. In 2020, 1.8 million cases were diagnosed in the United States [19]. However, as noted in the National Research Council report *Toward Precision Medicine: Building a Knowledge Network for Biomedical Research and a New Taxonomy of Disease* [20], cancer should no longer be considered a single disease, with different cancers distinguished by their location (e.g., breast or prostate), and degrees of progression (stage) and cell abnormality (grade). Molecular profiling or genomic sequencing of tumors, a key tool of precision oncology, has led to a new taxonomy of cancers and targeted treatments [21].

Cancers may be distinguished by their molecular profile. Identification of genomic alterations in an individual tumor may inform prognostication and treatment recommendations. Pharmacodiagnostics, involving the identification of molecular changes in a tumor, may predict whether a tumor will respond to a particular therapy [22]. In some cases, liquid biopsy, involving the collection and analysis of a body fluid sample to test for relevant biomarkers, may be used instead of a traditional invasive tissue biopsy [21,22]. Understanding tumor biology and the specific molecular changes in particular tumor types facilitates the development of therapies targeting tumors with particular molecular features.

Hormone therapies were among the earliest targeted therapies; they either prevent the production of hormones needed for cancer growth or prevent the hormones from acting on cells. Therapies may be targeted to specific features of a tumor's cells, for example, attacking deregulated proteins supporting cancer cell survival [22,23]. Some targeted therapies starve blood supply to tumors by inhibiting angiogenesis, thereby causing tumors to shrink or at least not grow. Stimulation of a patient's immune system to fight against cancer cells, immunotherapy, is another approach. Monoclonal antibodies, combined with chemotherapeutic agents, may be targeted to attach to cancer cells, delivering the agent to those cells while sparing noncancer cells. Just as cancer cells can develop resistance to traditional chemotherapy, resistance to targeted therapies can develop. Reprofiling tumors may be necessary, and the molecular profile of recurring cancer may differ from that of the original tumor.

Economic and policy implications

The short- and long-term social and economic implications of precision cancer interventions may differ. The cost of developing each targeted therapy is very high, particularly with research outcomes relevant only to a narrowly defined group of patients (defined by virtue of the molecular features of their tumors). In the long term, however, precision oncology may reduce healthcare costs by distinguishing those likely to respond to therapy from those for whom the therapeutic response is unlikely and offering the therapy only to likely-responders. If and when targeted cancer therapies are proven cost-effective, insurers are likely to cover them. Until then, high-cost targeted therapies are likely to exacerbate disparities in access to effective treatment and cancer-related morbidity and mortality. Moreover, because precision cancer treatments initially emerge in research-rich areas and more highly remunerative markets, geographic, informational, and educational barriers, as well as a lack of financial resources, currently prevent some patients from accessing them.

Some worry that developing therapies targeting narrowly defined groups will shrink developers' market shares and that developers will increase prices to offset reduced volume [24]. This would parallel economic challenges associated with treatment for rare diseases and could exacerbate cancer care and health disparities even in the long term. Nevertheless, it would be ethically valuable for tumor profiling to enable the targeting of therapies to those likely to benefit, while enabling others to avoid side effects of treatment that has a low (or no) probability of benefit. Benefitting patients, and reducing harm to them, is the ethical foundation of medicine.

Informed consent: understanding options and accessing treatment

Patients with decisional capacity have the right to give their informed, voluntary consent (or refusal) regarding recommended treatment options. (Those who lack decisional capacity should have a surrogate decision-maker weigh their options in light of their known values and personal interests, or the interests reasonably ascribed to them if their values and preferences are unknown.) Explaining complicated treatment options is frequently challenging for clinicians. Understanding those options is often harder for patients, and perhaps particularly difficult when they face a serious, even life-threatening diagnosis.

On the one hand, the possibility of tumor profiling and targeted treatment may narrow the range of options a patient needs to consider. On the other hand, information about rates of success of targeted treatment and prognostication will still be rife with probabilistic information. The public, patients, and even many clinicians struggle to understand and make decisions involving probabilities [25–27]. It may be especially challenging to help patients appreciate the probabilistic nature of information about prognosis and interventions when the treatment is termed precise, individualized, tailored, or targeted.

Moreover, like all of PM, precision oncology remains epidemiologically based. At least in some cases, targeted therapies that have a low probability of benefit for a particular patient, nevertheless cannot be said to have absolutely any possibility of benefit. Whether insurers will cover a "long shot" treatment when molecular profiling or genotyping reinforces its status as a "long shot" for a particular patient is a policy question raised by PM ever more narrowly defining the "indications" for particular therapies. Indeed, for many cancers, PM defines the condition in terms of its susceptibility to particular treatments.

This policy question is different from those focused solely on the high cost of individualized, targeted therapies. The question is instead about the right of patients with one molecular tumor profile to have access to treatment available to other patients if there is some chance that treatment may be somewhat effective. At least in the early days of targeted therapies, cancer patients were often required to exhaust all standard, "one-size-fits-all" treatments before insurance would cover tumor typing and targeted therapy [28]. As precision oncology develops more targeted treatments, insurers may use tumor typing outcomes to limit access to treatment that would previously have been available.

It may also be challenging to help patients understand the relationship between the genetic and molecular features of their tumor and the genetic changes, variants, or mutations that may have increased their risk of developing cancer. Cancer patients, like most people, lack the genetic literacy to understand the differences between somatic and germline testing [28]. Nevertheless, genomic profiling of a patient's tumor may reveal germline mutations that have implications for the patient's and the patient's family members' cancer risk and screening recommendations [29,30].

Incidental findings and informed consent

The discovery of a germline mutation related to hereditary cancer risk would be considered an incidental finding of tumor profiling undertaken to guide current cancer treatment. An incidental finding in the clinical context is defined as a finding of potential health or reproductive importance that is beyond or unrelated to the reason for conducting the test, examination, or analysis. Here the primary finding (the tumor's molecular profile) and the incidental finding (hereditary or familial cancer risk) are both related to cancer, but the latter isn't the primary reason for tumor profiling. Moreover, tumor profiling may also reveal germline variants associated with an increased risk for conditions other than cancer or reveal a genetic variant of uncertain significance [31].

The possibility of discovering incidental findings should be disclosed to patients undergoing tumor profiling (and indeed any genetic testing or genomic sequencing). Patients' preferences regarding whether and when they would want to learn such information—and whether and how they would share such information with family members—should be discussed. Some patients may find this additional information particularly unwelcome while dealing with a cancer diagnosis [32]. Employing techniques that only probe for genetic variants of immediate relevance to the patient's care (e.g., tumor-only sequencing) could reduce but not eliminate the possibility of incidental findings. Limiting the focus of sequencing and analysis may be cost-prohibitive or contrary to clinic, institutional, or health policies that embrace a practice of always seeking (at least clinically actionable) genetic variants [33,34]. Indeed, pairing somatic and germline analysis may become a common method of opportunistic screening for hereditary cancer risk [29].

Whether clinicians will defer to patients' preferences regarding such opportunistic screening or return of cancer- and noncancer-related incidental findings remains to be seen. Clinicians may be justifiably concerned about their legal liability if they receive health-relevant information from laboratories of which their patients remain unaware. For that reason, institutional policies may prohibit withholding such information from patients, even at the patients' request. Whether clinicians *should* defer to their patients' preferences depends in large measure on whether patients have both a right to-know and a right not-to-know health-relevant information about themselves, including genomic information.

The right to-know and not-to-know

The right to-know or to have access to one's own health information is a well-established right of adult patients in the United States. The legal and ethical right to give informed consent encompasses the right to have information about one's condition, as well as about the risks and potential benefits of recommended treatments so that one can make informed decisions about one's own care. The Health Insurance Portability and Accountability Act of 1996 (HIPAA) requires health plans and providers to give patients access to their health information upon request [35]. Access to such information is considered integral to their being able to direct their care, promote their health, monitor chronic conditions, ensure that information maintained about them is accurate, and generally promote and protect their well-being. Ethically, the right to-know is based on patients' right of self-determination or right to determine what happens to their bodies, and medicine's goal of promoting their well-being.

Having access to their health information enables patients to promote their own well-being and participate effectively in their health care.

Whether people have a right not-to-know information relevant to their health is somewhat more controversial. An important aspect of PM is identifying those who are at increased risk of developing disease—for example, those at increased cancer risk—and implementing targeted prevention strategies, such as surveillance and prophylactic interventions, including behavioral and lifestyle change. Knowledge of health-related risks is integral to this targeted prevention strategy.

Why might people not want to know such health-relevant information about themselves? In addition to the concern that they may be discriminated against or stigmatized based on their health risks or conditions (as discussed later), people may be concerned that learning of health risks will cause them anxiety or change the course of their lives. They may join the ranks of the "worried well" or become "previvors," survivors of an illness for which they have increased risk, but that they have not developed [36]. Knowing that one is specifically at increased risk for colon cancer, for example, maybe more burdensome than knowing general facts about colon cancer risk (e.g., that it is associated with a low-fiber, high-fat diet and lack of physical activity and is the fourth-most common cancer in the United States).

The magnitude of precision prevention's psychosocial burdens and potential benefits depends on people's values, circumstances, and priorities. One respondent in an interview about the value of genetic risk information commented that genetics "is not important. And it is not important because there is *so much else wrong*" [37, p. 64]. The "so much else wrong" may be a matter of personal challenges or structural social concerns, or both. The respondent was referring specifically to a lack of financial stability and basic health care, racism, and unemployment. People who face other pressing issues in their lives may not have the cognitive "bandwidth," time, and energy to make use of genetic risk information. For some, a lack of economic resources or discretionary time may prevent them from accessing preventive health care or engaging in health-promoting behaviors.

Qualitative research also indicates that learning individualized cancer risk information can have subtle psychological effects on individuals' self-concept, interpersonal and familial relationships, choice of career, the pace of seeking a mate, or timing of childbearing [38,39]. The timing of learning risk information matters both for mitigating negative psychosocial effects and maximizing the potential benefit of surveillance and prophylaxis.

As risk information becomes increasingly reliable, and if preventive interventions become increasingly effective and accessible, and less onerous, it is plausible to think that the benefit of receiving such risk information will more frequently outweigh the benefit of forgoing it. In other words, as learning risk information becomes less practically, economically, and psychically burdensome because preventive interventions become easier and more clearly beneficial, an assertion of the right not-to-know may dissipate. Nevertheless, preventive health care may never become a priority for some people due to their personal values and view of a good life.

A duty to warn?

The question of whether people have the right to-know or not-to-know of their health risks has always accompanied discussion of whether there is a right and/or obligation to inform patients' family members of genetic risks of disease. Is there a "duty to warn" genetically related family members? If so, who has that duty: patients or clinicians? Do patients have an ethical obligation to share information relevant to their relatives' health? Do clinicians have an ethical or legal obligation to inform third parties of information about their patients to help those nonpatient third parties avoid harm? May clinicians violate the duty of confidentiality owed their patients to fulfill a "duty to warn" third parties?

Although cancer care is a particularly relevant context in which to consider these questions, the legal question of a clinician's duty to warn nonpatient others arose most famously in the context of psychiatry and has prompted decades of ethical debate. There, in *Tarasoff v. Regents of the University of California*, it was held that notwithstanding their duties to protect patient confidentiality, mental health professionals have a legal "duty to protect" identifiable third parties when a patient presents a credible risk of serious avoidable bodily harm to those others [40]. Warning them so that they may take measures to protect themselves is one way to discharge this duty, a duty that arises, according to *Tarasoff*, when three conditions are satisfied: (1) the risk is credible, (2) it is a risk of serious and avoidable/preventable harm, and (3) the risk is presented to specific identifiable others. Though this was a legal decision in the context of mental health care, many commentators consider these considerations relevant to determining, in a wide range of healthcare contexts, whether there is not only an ethical obligation for clinicians to warn others but also a right to breach their patients' confidence to do so.

How do these conditions apply in the context currently most relevant to PM—namely, with regard to genetic risks? That the risk be credible would require, at minimum, that the information (e.g., test result) is reliable and valid. A small increased risk of a very serious harm (like a tiny risk of a life-threatening anaphylactic reaction) may be given more psychological weight (one meaning of "credibility") than a high risk of minor harm (as when new kindergarten teachers have a high risk of catching colds from their young charges). For genetic relatives of patients with a germline mutation increasing their risk for cancer, there is often a 50/50 chance that they have inherited the same mutation. It is a matter of personal values whether this chance would constitute a credible risk, even if everyone would consider having cancer a serious harm. It has been argued that since the mutation is either already present or not, it is not avoidable; however, developing cancer can in some cases be prevented or early treatment can mitigate its harm. Whether those potentially at risk—namely, the patient's genetic relatives—are "identifiable" by a clinician depends on circumstances and the patient's cooperation.

Hereditary cancer risk was at issue in the two often-cited legal cases considering whether there is an obligation on the part of a clinician to warn a patient's family members of the possibility of increased genetic risk. In the Florida case of *Pate v. Threlkel*, the court found that a clinician does have a duty to warn a patient's immediate genetically related family members, but held that this duty was fulfilled by informing the patient of the hereditary or genetically transmissible nature of the disease risk [41]. In contrast, the New Jersey case of *Safer v. Pack* held that informing only the patient might not satisfy a clinician's duty to warn those known to be at risk of avoidable harm from a genetically transmissible condition [42]. The court's decision in *Safer* left it to future jury deliberations to determine whether clinicians have a duty to inform relatives directly.

The law in the United States has remained in flux and at odds on this question of whether clinicians have a duty to warn patients' family members directly. The consensus seems to hold that clinicians should inform patients of the familial relevance of health risk information (genetic or otherwise) and should then support patients' efforts to share that information with relatives, but that clinicians ought not to undertake to contact relatives themselves without their patient's permission. The consensus rests on the practical impossibility of identifying and contacting family members without the patient's cooperation, as well as the ethical value of maintaining patient confidentiality and the material value of maintaining patients' trust to sustain a therapeutic alliance.

Intrafamilial risk disclosure

Patients should consider whether they have an ethical obligation to share with family members the health risk information that they learn. Again, the example of hereditary cancer risk provides a classic example; however, PM should eventually involve identifying and reducing environmental, behavioral, lifestyle, and cultural risk factors. Lead exposure risk, or risk of other environmental toxins, for example, might be subject to the same analysis as hereditary risks of cancer or other diseases (like diabetes or cardiovascular conditions). Family members may be exposed to the same environmental risk factors or may have a similar lifestyle.

Some have argued that individuals have a duty to warn or a "duty to rescue" family members from the genetic risk of harm due to the familial nature of some genetic risks, the special nature of familial obligations, and/or a general duty to help those in need [43]. Others respond that the magnitude and probability of such genetic risk rarely rise to the level generally thought to trigger a duty to "rescue," similar to the duty arising when one passes a drowning child or can prevent someone from stepping into traffic [44]. They argue that potential benefits to the "rescuees" (family members) may appropriately be balanced against psychosocial and other costs to the "rescuer" (patients) [45]. They also note that "rescuees" typically have alternate means of "rescue," that is, unlike a drowning child who likely cannot rescue herself, individuals may seek genetic testing/screening [44]. Particularly as PM evolves and information about health-related risks to some subpopulations becomes increasingly widely available, individuals may themselves undertake to learn whether they fall into those risk categories or are members of risk-defined populations.

In addition, as PM progresses, research on people's preferences regarding learning risk information, and even more important, study of whether and how they actually act on risk information, may inform both patients' and clinicians' deliberations about sharing health-related risk information [46−49]. Although people's preferences are not decisive in determining whether and what health-related information should be supplied to them, their preferences are relevant to decisions and policies about information sharing [50].

Informed consent in precision oncology research

As for all of PM, the knowledge base for precision oncology is still developing. When individualized, molecular profile-guided treatment is offered in the context of clinical research, the likelihood of patients experiencing therapeutic misconception would seem to be increased. Therapeutic misconception involves patient-participants mistakenly believing that the purpose of the research study is to provide treatment, that it will (or is designed to) benefit them, or that the intervention they receive will be chosen to benefit them individually [51,52]. Describing a study intervention as precise, personalized, individualized, or targeted may make it more difficult for prospective participants to appreciate both the nonindividualized, protocolized nature of research designed to test such therapies and the increased uncertainties involved while the therapies are still in development. Particularly in the context of cancer—a condition that leads patients to experience stress, time pressure, a mix of reasonable and unrealistic hope, and even desperation—it may be challenging to convey the concept of *protocolized research* on *individualized treatment*.

During the consent process for research participation, particular disclosures must be made. These include emphasizing that what the participant is being asked to do is participate in *research*, discussion of alternatives to study participation, and explanation of what is known and unknown regarding the pathophysiology of the disease condition, the function of the particular targeted therapy, and risks associated with the therapy (e.g., that immune checkpoint inhibitors trigger autoimmunity with immune-related adverse events) [53]. Especially for people with cancer or other serious illnesses, it may be important to acknowledge that some research opportunities and treatment options may cause them more harm than the potential for benefit and that participants' values and considerations of quality of life matter.

Finally, in research as in clinical care, the possibility of incidental findings must be discussed during the informed consent process, especially for studies involving genomic sequencing. Among the models for informed consent and management of incidental findings is a "traditional consent model" whereby consent to return such findings is obtained at the time of study enrollment [54]. A "staged consent model" involves the disclosure of the possibility of incidental finding discovery during the consent process for participation. More detailed consent is then obtained if such findings are discovered. The future attempt to obtain consent for the return of a particular incidental finding, however, will necessarily disclose the finding to some degree. If participants do not consent to receive full information about the incidental finding, they may misunderstand the partially disclosed information. Finally, requiring that participants consent to "mandatory return" of some set or class of incidental findings as a condition of study enrollment may be contrary to the voluntariness requirement of informed consent, may introduce bias in the study population, and may not serve all participants' interests.

Polygenic risk assessment and complex conditions: health disparities, disparate benefits of risk assessment, and the risks of discrimination, stigmatization, and blame

Beyond cancer, PM seeks to use targeted strategies to prevent other common complex conditions resulting from multiple genomic variants and environmental influences. Polygenic risk assessment, a tool of precision prevention, seeks to target the right preventive interventions, at the right time, to the individuals who need them most in order to prevent the onset of disease. Though genetic risk analysis is not PM's only tool of prevention, it is currently the most prominent. After substantial investment in genome-wide association studies to identify genetic variants associated with disease, there is now substantial investment in translating those findings into polygenic risk assessment and precision prevention, though thus far the clinical utility of such assessment has been less robust than anticipated [55–57].

Of course, other risk assessments may focus on the environmental factors contributing to disease manifestation. Many nongenetic, environmental risk assessments can lead to either population-wide or targeted interventions, or both. These include testing the lead level in community water supplies and replacing lead pipes or the source of water; testing community or workplace air quality and implementing restrictions on pollutants or improved ventilation systems; regular testing of food supply chains to ensure safety or require improvements; and occupation- or housing-related health and safety hazard assessment with subsequent remediation. Nevertheless, polygenic risk assessment is garnering substantial scientific and ethical attention.

Although polygenic risk assessment is only beginning to enter clinical practice, direct-to-consumer genetic testing companies will report a polygenic risk score (PRS) for various complex conditions, including diabetes and cardiovascular conditions [58]. Currently, it is wise to question the accuracy of DTC test results before making important health-related decisions, as the quality of both laboratory testing and the reference genomes used for interpretation may vary [59]. Even when accurate, most PRSs are only "moderately predictive of group outcomes," and even those that are accurate and highly predictive would apply to groups, but not necessarily predict an individual's personal disease risk or health outcome [60].

A PRS is the result of statistical analysis. Because most people, and even clinicians, lack statistical literacy [25−27], it may be particularly difficult for individual patients to appreciate the relevance of their relative risk for disease represented by a PRS. Patients are typically concerned about their absolute risk of developing a condition. Sometimes they are concerned to know their risk of developing a condition at any point in their lifetime, but sometimes it is their risk of developing a condition during a particular time of life that is of interest. Moreover, one's absolute risk of developing a disease may be affected by preventive interventions or changing environmental factors. The evolving state of science with newly emerging information about risk factors and an improved understanding of disease etiology may alter risk assessments, and a degree of "irreducible unpredictability" always accompanies risk assessment [61].

Health disparities and disparate benefits from precision medicine

Most problematic is the fact that genomic studies have overwhelmingly examined genomes of people of European ancestry. There is insufficient data about genomic variants in populations of other ancestries to accurately calculate a PRS regarding most conditions for individuals of those ancestral backgrounds [62,63]. That polygenic risk assessment is currently mostly valid and useful only for those of European ancestry is likely to worsen health and healthcare disparities [62]. If these unjust disparities are to be addressed, it is critical to enroll diverse and representative populations in genomic research [64,65].

To achieve the promise of PM, however, research needs to enroll participants not only with diverse continental ancestries but also from diverse socioeconomic and environmental contexts. If PM is to reduce rather than exacerbate health and healthcare disparities, "the *relative importance* of bias, racial discrimination, culture, socioeconomic status, access to care, environmental factors, and genetics to racial/ethnic differences in disease" must also be studied further [66, p. 478−9, emphasis added]. Studying risk factors for conditions particularly prevalent in underrepresented and underserved populations is also important to reduce health disparities [62].

The All of Us Research Program, a program of the National Institutes of Health that grew out of President Obama's Precision Medicine

Initiative, seeks to enroll participants from diverse continental ancestries and to collect not only health history and genomic (and other omic) data but also a wide range of information about lifestyle, behaviors, and environmental exposures through self-reporting and wearable devices [67,68]. It promises to be a valuable research platform for the development of the knowledge base of PM. Because of the potential diversity of a million American research cohort, the Program is considered one of the most promising of all the global biobanking projects [37].

Currently, the clinical use of polygenic risk assessment may be premature for many people and with regard to many conditions, and not only because of concerns about its accuracy and generalizability. Studies are prompting skepticism that supplying risk information actually leads people to engage in the sustained behavioral changes that are usually necessary to prevent disease [69,70]. To date, "expectations that communicating DNA-based risk estimates changes behavior is [sic] not supported by existing evidence" [46, p. 1].

There may be multiple barriers to acting on health-related risk information, including financial, cultural, circumstantial, educational, and religious/values-based barriers. Socioeconomic barriers are of particular concern from the perspective of social justice [71]. While risk information should not be withheld from those who are likely to lack socioeconomic resources to act on that information, it should be acknowledged that health-related risk information may not afford them the same potential benefit that it offers those who have greater wealth, education, or cognitive and affective "bandwidth" to make use of it [72]. Remember the study interviewee who commented that genetic risk information did not matter because so much else was wrong [37]. Insofar as there are groups who disproportionately face "so much else" that serve as barriers to acting on health risk information—and especially if those barriers result from systemic injustice—the difference in potential benefit from a risk assessment is unjust and requires attention.

Discrimination, stigmatization, and blame

PM is not supposed to group people solely on the basis of their genotypes; their environmental exposures, lifestyle, and specific behaviors also contribute to risk assessment and to their categorization into different subpopulations that are at particularly increased risk for disease or that could benefit from targeted preventive or, if necessary, therapeutic interventions.

Insofar as behavioral and lifestyle differences—and even some environmental exposures—may be considered a matter of choice, there is a risk that PM may increase the tendency to blame people for incurring those health risks or developing their conditions.

Individualizing risk—either by calculating a person's individual PRS or by focusing on the contribution of their individual behaviors to their health-related risk—encourages the assignment of individual responsibility for ill health. Once the association between smoking and lung cancer was established, there has been a tendency to blame lung cancer patients who smoked for causing their cancer, even though it is likely that their genetics also plays a role in whether they develop cancer. People whose weight is unhealthy for them are frequently blamed and stigmatized for being obese, even though genetic factors outside of their control likely contribute to their metabolism and other physiological factors related to their weight.

Currently, in PM, the environmental component of gene x environment interaction is frequently understood in largely individual terms. This understanding is reflected in research that gathers data from wearable technologies and asks about health behaviors, but does not sample air and water quality or include the study of the physiological impact of stresses of racism, violence, food insecurity, or income inequality. PM may benefit from the integration of more behavioral research but should resist stopping its analysis or interventions at individuals' behaviors. In studying weight and developing interventions to reduce weight-related risk of developing diabetes or cardiovascular disease, for example, behavioral research should not focus solely on diet and exercise, but should also examine social structures, environments, and social influences on ego depletion that may underlie dietary and exercise behaviors. Individualizing "environment"—that is, focusing primarily on behavioral or lifestyle factors—distracts from attending to natural, social, and built environmental factors such as pollution, plastics, and workplace risks, as well as infrastructural factors that affect lifestyle "choices" that are the subject of economic and behavioral economic studies [73].

Even while using polygenic risk assessment, it is important for precision prevention not to focus solely on individual-level interventions and personal behavior change. There are not only socioeconomic barriers to behavioral change that require redress but also important changes in a physical and social environment that would directly improve individual and population health.

Finally, it has long been recognized that if individuals are to make use of their individual health-risk information—specifically, genetic risk information—there must be protections from discrimination on the basis of those risks. Under the Genetic Information Nondiscrimination Act (2008) or GINA, discrimination in employment and health insurance underwriting based on genetic information is largely prohibited [74]. Some of those who are not covered by GINA—e.g., federal and military employees and people who receive their health care through the military, Veterans Administration, Indian Health Service, or Federal Employee Health Benefits Plan—have similar protections through other policies or by Executive Order. However, employees in settings with fewer than fifteen employees do not have such protections. Furthermore, there is no prohibition on using genetic risk information in life, disability, and long-term care insurance underwriting to deny insurance or charge a higher rate for it. Nor can social attitudes be legislated. Genetic risk for ill health may lead to social stigmatization.

Pharmacogenomics

While polygenic risk assessment is only beginning to be implemented in clinical care, PGx is already experiencing widespread clinical translation in addition to its research applications. Some of the ethical considerations arising in other domains of PM apply to PGx.

Pharmacogenomics research and drug development

Pharmacogenomic research studies genetic variations that influence a person's response to a drug or class of drugs. The use of PGx testing in research and drug development increases the likelihood and speed of drug approvals. Drug developers may "rescue" and bring to market a drug that is ineffective for a study population as a whole if it proves to be effective for a subset of study participants. By characterizing that subset by genotype, and then enrolling study participants with that genotype in smaller follow-up studies, drug developers are able to seek approval and subsequently market the resulting drug for patients with that genotype. Approvals may be faster, and the costs and risks of clinical trials and drug development may be reduced [75]. However, the development of policies and/or financial incentives may be necessary to ensure that patients with other condition + genotypes are not "orphaned" by the possibility of developing drugs to treat subsets of study populations while ignoring other subsets of patients with the same condition [76].

In addition, enrolling participants in clinical studies based on their genotype—either for drug development or to study disease—presents complex informational, psychological, and thus ethical challenges [77]. Prospective participants may not be aware that they have a genotype that makes them eligible and attractive for enrollment in a particular study. Whenever individuals are recruited based on their genotype, questions arise about how to approach them for enrollment without imposing unwanted information. To explain why they are being approached, or in explaining the purpose of the study, they may become aware of the possible (or established) significance of their genotype [78]. That information may or may not be welcome. When the information is fairly innocuous—for example, a genotype associated with slow or rapid drug metabolism—informing people "out of the blue" in the course of study recruitment may be relatively unproblematic. If the information is instead connected to disease risk, or if the drug under study is related to a particular disease condition, learning the information may not be psychosocially or ethically benign.

Clinical use of pharmacogenomics

In the clinical context, PGx involves using this information to guide prescribing drugs to individual patients at appropriate dosages. Differences in drug response are known to be caused by differences, for example, in the metabolism of particular agents, or the transport or rate of uptake into particular targets (e.g., specific tissues or cells); different genotypes are associated with these physiological differences between people. Those who metabolize a drug more rapidly, for example, may need a higher or more frequent dose for it to be effective. Those who metabolize it more slowly may need less and with usual doses may have a higher risk of adverse drug reactions (ADR).

What could be ethically problematic in using genetic testing or genomic sequencing to reduce the risk of patients having ADR? What could be wrong with increasing the likelihood that pharmacologic treatment of their conditions is effective, thereby reducing healthcare costs and patient suffering, both in general and for individual patients? Unlike information risk of disease, there is nothing particularly stigmatizing or psychosocially disruptive, for example, about being a slow or rapid drug metabolizer.

However, because a single genotype may be associated with multiple phenotypes, a phenomenon called "pleiotropy," genetic variants associated with drug response may also be associated with disease risk. The possibility that PGx testing may yield an incidental finding of disease risk may be a reason to obtain specific informed consent for PGx testing, or at least to inform patients of this possibility prior to PGx testing. Whether a patient should be permitted to refuse PGx testing to avoid learning disease risk information and still be prescribed the drug for which testing is recommended is an open ethical question, and one that likely is situation specific. The answer would likely depend on the specific risks involved, the costs and challenges of monitoring for ADR, the magnitude and probability of the therapeutic benefit from available alternatives, and the patients' reasons, values, and circumstances.

Patients should also be informed of the potential familial relevance of PGx test results and incidental findings. While it may seem obvious that a patient should share PGx test results with family members because they may have inherited the same genetic variants, it may be impossible to share PGx information without sharing other genetic risk information. Because the interpretation of genetic variants changes over time, learning and sharing a PGx test result may at some future date amount to having learned or shared disease risk information [79].

Except for these concerns about incidental findings and intrafamilial sharing of genetic information, clinical use of PGx test results may seem deceptively straightforward. Simplistically described, PGx testing sorts people into groups based on genotypes, which are associated with cellular-level responses to a drug, which are in turn associated with effective treatment and/or ADR. Those at low risk of ADR with a high likelihood of benefit should be prescribed the drug. Those at high risk of ADR with a low likelihood of benefit should not. But what if there is no safer or more effective alternative? In that case, prescribing the drug and carefully monitoring for ADR may be appropriate.

However, PGx testing groups patients according to probabilities of ADR and therapeutic benefit, rather than identifying with certainty which individuals will/won't benefit or which will suffer adverse reactions. Thus, there may be reasons to prescribe contrary to PGx test result indications. These reasons include patient circumstances and preferences, drug costs and convenience, and insurance coverage. As in other clinical medicine contexts where patients are permitted to make decisions in light of their values, circumstances, and degree of risk aversion or proneness, it may be appropriate to allow patients to accept associated risks of treatment in hope of receiving benefits, though permitting them to do so raises considerations of liability for any ADR (discussed later).

Moreover, in many cases, PGx testing does not yield a prescribe/don't prescribe recommendation. Instead, PGx testing is used to determine the dosing regimen of specific drugs or to guide monitoring for ADR. Slow metabolizers, for example, may need more careful monitoring for ADR while identifying how small a dose will still have therapeutic benefits. Usual considerations in applying dosing guidelines to a particular patient are relevant; clinical judgment is required, for example, when applying PGx-informed dosing guidelines for adults to pediatric patients [80].

As is true with clinical practice guidelines and clinical decision support in general, recommendations to use PGx testing to guide dosing may impose an obligation (ethical, legal, or both) for clinicians to employ PGx testing, once such testing is considered part of standard care. How insurers will and should regard such indications and prescribing practices remains an open question. While GINA prohibits health insurers from denying people coverage or charging them higher rates based on genetic information, insurers are permitted to use genetic information to determine medical needs and, in turn, what is "medically necessary" treatment [81]. A strong family history of breast cancer or positive BRCA1/2 testing for a pathogenic variant, for example, may be used to establish a young woman's medical need for mammography that would not otherwise be covered by insurance until she is older. Particularly as PGx testing is more frequently recommended before prescribing a drug, it remains to be seen whether and under what conditions insurers deny coverage for patients with genotypes indicating an increased risk for ADR or a lower probability of benefit.

Development of clinical guidelines and development of health insurance regulations or guidelines should take into account patients' medical needs and the growing body of PGx information. Consideration of patient well-being, as well as healthcare cost containment and just allocation of resources, supports using PGx testing. Nevertheless, because PGx yields results applicable with confidence to patient groups, but not with certainty to individual patients, concern for well-being and fairness suggests that an individual whose genotype indicates a relatively lower probability of positive treatment response should not be categorically denied access to (or coverage for) that intervention so long as three conditions are met. These conditions are as follows: (1) The intervention is otherwise available/covered; (2) no other intervention is available that is likely effective for patients with that genetic variation; and (3) the risk or burden of the intervention to the individual—that is, the probability and magnitude of ADR—is not unduly burdensome given the magnitude and probability of benefit. With regard to (3), some bioethicists have argued that an individual should be allowed to assume those risks and not be denied the opportunity to seek even remote benefits when no other good alternative exists and when others with a different genotype have access to the intervention [82].

Who should bear the liability for ADR (and financial responsibility for treating sequelae of ADR), if a drug is prescribed contrary to a PGx test result, is unclear. In the absence of an alternative—or if alternatives are not affordable or are otherwise unattractive—a patient may want to assume the risk of experiencing ADR and receive the "contraindicated" drug. If the intervention or drug with a remote (or relatively lower) prospect of benefit could be administered at the patient's own expense, it would seem quite reasonable to allow the patient to assume ADR-related risks of the personally preferred drug. But, most patients cannot afford to pay for medication or chemotherapy, or for treatment of ADR, themselves. Therefore, patients' requests for access to "genotype-contraindicated" treatment have relevance for healthcare institutions, insurers, and members of insurance plans whose rates are partially dependent on what insurers are required to pay. It is reasonable for healthcare providers to consider these downstream costs when considering exceptions to clinical guidelines, and for insurers to consider them when adjudicating coverage appeals.

Nevertheless, genotype-based decisions to deny an individual an intervention that is available to others should not be made lightly or categorically. Appeal processes should be established, or existing appeals processes should be expanded, to address denials based on PGx results, just as they were for denials based on clinical guidelines informed by comparative effectiveness research findings [83]. The sorting of people into different groups that is intrinsic to PM should not be used to disadvantage people who would be afforded a chance to benefit from treatment in the absence of such sorting.

It may be appropriate to afford patients a right to try a "genotype-contraindicated" drug, rather than permitting PGx testing to "orphan" a group of patients with a condition + genotype that has no available treatment, while there is a treatment for other patients with the same condition. Though such requests might be modeled on "right to try" requests in the context of investigational drugs, here the drugs would be part of the standard of care. Unlike the "right to try" in the context of investigational interventions, here there would be no ethical concerns about such requests undermining a clinical trial's production of generalizable knowledge [84]. Further, measures may be developed to protect clinicians and institutions from liability for ADR, similar to the protections afforded by "right to try" provisions in the investigational drug context where manufacturers are protected from liability for adverse outcomes.

Expanded need for genetic counselors throughout precision medicine

These considerations—regarding PGx testing and individuals' access to drugs that have a poor risk:benefit profile for them—are also relevant to explaining and accommodating patients' preferences regarding individualized cancer therapies. Indeed, in all domains of PM, patients would likely benefit from the skills and knowledge of genetic counselors to help them understand the risks they are accepting. Genetic counseling could help patients appreciate the familial relevance of test results or polygenic risk assessment and then navigate intrafamilial sharing of such information. Unfortunately, in many clinical settings where PGx testing is performed, genetic counseling may not be available. Genetic counselors are more likely to be involved in rare disease and precision oncology settings. In primary care settings, however, where PRSs are ultimately going to be used to guide preventive health care, referral for genetic counseling is constrained by the current shortage of genetic counselors.

Conclusion

This chapter outlined many of the ethical and policy considerations associated with PM. The issues discussed arise in multiple domains of PM, though they were presented in relation to one or another domain: PGx, polygenic risk assessment, precision oncology, or diagnosis and treatment of rare diseases. In each context, the chapter sought to illuminate how values influenced the decisions to be made.

Individual patients' values, and their priorities in light of their circumstances, influence their decisions to-know or not-to-know about their genetic and other health-related risks, as well as whether to share that information with their family members. Through the informed consent process, patients are asked to make decisions to accept or refuse recommended healthcare interventions based on their own values and circumstances. For some of those who are ill, PM may provide better, more specific information to inform those decisions. Genetic/genomic testing may improve diagnosis and end diagnostic odysseys for some, particularly those with rare conditions. PGx will guide prescribing and dosing for many conditions. In cancer care, PM may enable prevention, earlier diagnosis, and more effective treatment.

Because of the intensified preventive approach of PM, people will increasingly be asked to make decisions prior to disease onset, including whether to learn their specific risks of disease and whether to engage in prophylactic interventions. PM's preventive recommendations will be more population-specific than previous eras' generalized preventive health advice. Based on their values and circumstances, people may differ regarding how they value preventive information, including information about their disease risk based on their membership in particular subpopulations.

In addition to these value-based decisions to be made by individuals, social entities—healthcare providers, insurers, courts and juries, policymaking bodies, and professional societies—will continue to make value-laden decisions as the knowledge base for PM grows. These decisions include determining what probability and magnitude of risk warrants a recommendation to initiate a preventive intervention, and determining that a therapeutic intervention has a sufficient prospect of benefit so that it should be approved by regulators, covered by health insurance, or prescribed to a particular patient.

Values inform decisions about which health risks are sufficiently credible and serious, as well as reasonably preventable so policymakers or courts might justifiably impose an obligation on clinicians or individual patients to warn patients' family members. Similar values will inform whether people continue to have a right not-to-know information (particularly genetic information) about their health risks. Values—regarding fairness and well-being—will inform decisions about which social institutions and practices are not permitted to discriminate on the basis of genetic or other health information (e.g., employers and health insurers) and which may do so (e.g., life insurers, people choosing friends, or marriage partners).

Behind the scenes, before clinicians interact with individual patients, and even before policies are determined, values will have played a role in the science that creates the knowledge base for PM. Values influence the determination of which populations to study, what constitutes a representative study population, and which characteristics and data are used to sort people into subpopulations [62,85,86]. Values are involved in evaluating evidence to interpret genetic variants (as benign, pathogenic, or of uncertain significance) and in judging evidence to be of sufficient quality to be clinically relevant [87,88]. Policymakers and research funders make value-laden decisions to allocate limited funds across genomic, environmental, and behavioral health research programs.

Though these "behind the scenes" values may not be ethical values per se, they reflect views about what is fair, what is important for individuals' well-being, and what decisions should be left to individual choice, made by experts, or made at a societal level through social deliberation or political processes. Individuals—as voters, taxpayers, patients, and caregivers—have a stake in these value-laden decisions that range from broad policymaking to genetic variant interpretation, or from research funding decisions to the development of clinical guidelines.

References

[1] Obama B. Remarks by the President on precision medicine, from https://obamawhitehouse.archives.gov/precision-medicine; 2015, January 30 [retrieved 01.02.23].

[2] Juengst E, McGowan ML, Fishman JR, Settersten Jr. RA. From "personalized" to "precision" medicine: the ethical and social implications of rhetorical reform in genomic medicine. Hastings Cent Rep [Internet] 2016;46(5):21—33. Available from: https://doi.org/10.1002/hast.614.

[3] President's Council of Advisors on Science and Technology. Priorities for personalized medicine. Available from https://bigdatawg.nist.gov/pcast_personalized_medicine_-priorities.pdf; 2008.

[4] Tabor HK, Goldenberg A. What precision medicine can learn from rare genetic disease research and translation. AMA J Ethics [Internet] 2018;20(9):E834—40. Available from: https://doi.org/10.1001/amajethics.2018.834.

[5] Ball P. Bursting the genomics bubble. Nat [Internet] 2010;. Available from: https://doi.org/10.1038/news.2010.145.

[6] Basel D, McCarrier J. Ending a diagnostic odyssey: family education, counseling, and response to eventual diagnosis. Pediatr Clin North Am [Internet] 2017;64(1):265—72. Available from: https://doi.org/10.1016/j.pcl.2016.08.017.

[7] Lazaridis KN, Schahl KA, Cousin MA, Babovic-Vuksanovic D, Riegert-Johnson DL, Gavrilova RH, et al. Outcome of whole exome sequencing for diagnostic odyssey cases of an individualized medicine clinic: the Mayo Clinic experience. Mayo Clin Proc [Internet] 2016;91(3):297—307. Available from: https://doi.org/10.1016/j.mayocp.2015.12.018.

[8] Michael W. Zebras do exist: the diagnostic odyssey of rare-disease patients. 2020 https://rarepatientvoice.com/wp-content/uploads/2020/09/Zebras-Do-Exist-the-Diagnostic-Oddysey-of-Rare-Patients.pdf.

[9] Nguyen MT, Charlebois K. The clinical utility of whole-exome sequencing in the context of rare diseases — the changing tides of medical practice: The clinical utility of

whole-exome sequencing. Clin Genet [Internet] 2015;88(4):313—19. Available from: https://doi.org/10.1111/cge.12546.

[10] Pearson H. Human genetics: one gene, twenty years. Nat [Internet] 2009;460 (7252):164—9. Available from: https://doi.org/10.1038/460164a.

[11] Rosell AM, Pena LDM, Schoch K, Spillmann R, Sullivan J, Hooper SR, et al. Not the end of the odyssey: parental perceptions of whole exome sequencing (WES) in pediatric undiagnosed disorders. J Genet Couns [Internet] 2016;25(5):1019—31. Available from: https://doi.org/10.1007/s10897-016-9933-1.

[12] Thorp E. The emotional odyssey of a rare disease diagnosis: digging deeper, to do better [Internet]. J mHealth 2022;. Available from: https://thejournalofmhealth.com/ the-emotional-odyssey-of-a-rare-disease-diagnosis-digging-deeper-to-do-better/.

[13] Wu AC, McMahon P, Lu C. Ending the diagnostic odyssey-is whole-genome sequencing the answer? JAMA Pediatr [Internet] 2020;174(9):821—2. Available from: https://doi.org/10.1001/jamapediatrics.2020.1522.

[14] United States Congress. Orphan Drug Act of 1983. Pub L No. 97—414, 96 Stat 2049—2057. https://www.govinfo.gov/content/pkg/statute-96/pdf/statute-96-pg2049. pdf; 1983.

[15] Capra E, Reichert M, Salazar P. Precision medicine in practice: strategies for rare cancers [Internet]. Mckinseycom McKinsey & Co; 2021. Available from: https:// www.mckinsey.com/industries/life-sciences/our-insights/precision-medicine-in-practice-strategies-for-rare-cancers.

[16] National Organization for Rare Disorders. Updated rare disease facts and figures from NORD—rare & undiagnosed network [Internet]. Available from: https://rareundiagnosed.org/rare-disease-facts/; 2017.

[17] Nolen S. Nations face crushing cost of medical miracles. New York Times 2023;A1 January 23.

[18] Schlander M, Beck M. Expensive drugs for rare disorders: to treat or not to treat? The case of enzyme replacement therapy for mucopolysaccharidosis VI. Curr Med Res Opin [Internet] 2009;25(5):1285—93. Available from: https://doi.org/10.1185/ 03007990902892633.

[19] National Cancer Institute. Cancer statistics [Internet]. https://www.cancer.gov/ about-cancer/understanding/statistics; 2015.

[20] National Academies Press. Toward precision medicine: building a knowledge network for biomedical research and a new taxonomy of disease. US: National Academies Press; 2011.

[21] Schwartzberg L, Kim ES, Liu D, Schrag D. Precision oncology: who, how, what, when, and when not? Am Soc Clin Oncol Educ Book [Internet] 2017;37:160—9. Available from: https://doi.org/10.1200/EDBK_174176.

[22] Fernandez-Rozadilla C, Simões AR, Lleonart ME, Carnero A, Carracedo Á. Tumor profiling at the service of cancer therapy. Front Oncol [Internet] 2020;10:595613. Available from: https://doi.org/10.3389/fonc.2020.595613.

[23] Lee YT, Tan YJ, Oon CE. Molecular targeted therapy: treating cancer with specificity. Eur J Pharmacol [Internet] 2018;834:188—96. Available from: https://doi.org/ 10.1016/j.ejphar.2018.07.034.

[24] Krzyszczyk P, Acevedo A, Davidoff EJ, Timmins LM, Marrero-Berrios I, Patel M, et al. The growing role of precision and personalized medicine for cancer treatment. Technol (Singap World Sci) [Internet] 2018;6(3—4):79—100. Available from: https://doi.org/10.1142/S2339547818300020.

[25] Gigerenzer G, Gaissmaier W, Kurz-Milcke E, Schwartz LM, Woloshin S. Helping doctors and patients make sense of health statistics. Psychol Sci Public Interest [Internet] 2007;8(2):53—96. Available from: https://doi.org/10.1111/j.1539-6053.2008.00033.x.

[26] Martyn C. Risky business: doctors' understanding of statistics. BMJ [Internet] 2014;349(sep17 4):g5619. Available from: https://doi.org/10.1136/bmj.g5619.

[27] Schmidt FM, Zottmann JM, Sailer M, Fischer MR, Berndt M. Statistical literacy and scientific reasoning & argumentation in physicians. GMS J Med Educ [Internet] 2021;38(4):Doc77. Available from: https://doi.org/10.3205/zma001473.

[28] McGowan ML, Settersten Jr RA, Juengst ET, Fishman JR. Integrating genomics into clinical oncology: ethical and social challenges from proponents of personalized medicine. Urol Oncol [Internet] 2014;32(2):187−92. Available from: https://doi. org/10.1016/j.urolonc.2013.10.009.

[29] Forman A, Sotelo J. Tumor-based genetic testing and familial cancer risk. Cold Spring Harb Perspect Med [Internet] 2020;10(8):a036590. Available from: https:// doi.org/10.1101/cshperspect.a036590.

[30] Jain R, Savage MJ, Forman AD, Mukherji R, Hall MJ. The relevance of hereditary cancer risks to precision oncology: what should providers consider when conducting tumor genomic profiling? J Natl Compr Canc Netw [Internet] 2016;14(6):795−806. Available from: https://doi.org/10.6004/jnccn.2016.0080.

[31] Yushak ML, Han G, Bouberhan S, Epstein L, DiGiovanna MP, Mougalian SS, et al. Patient preferences regarding incidental genomic findings discovered during tumor profiling: patient preferences and tumor profiling. Cancer [Internet] 2016;122 (10):1588−97. Available from: https://doi.org/10.1002/cncr.29951.

[32] Hamilton JG, Shuk E, Genoff MC, Rodríguez VM, Hay JL, Offit K, et al. Interest and attitudes of patients with advanced cancer with regard to secondary germline findings from tumor genomic profiling. J Oncol Pract [Internet] 2017;13(7): e590−601. Available from: https://doi.org/10.1200/jop.2016.020057.

[33] Green RC, Berg JS, Grody WW, Kalia SS, Korf BR, Martin CL, et al. ACMG recommendations for reporting of incidental findings in clinical exome and genome sequencing. Genet Med [Internet] 2013;15(7):565−74. Available from: https://doi. org/10.1038/gim.2013.73.

[34] Miller DT, Lee K, Gordon AS, Amendola LM, Adelman K, Bale SJ, et al. Recommendations for reporting of secondary findings in clinical exome and genome sequencing, 2021 update: a policy statement of the American College of Medical Genetics and Genomics (ACMG. Genet Med [Internet] 2021;23(8):1391−8. Available from: https://doi.org/10.1038/s41436-021-01171-4.

[35] United States Congress. Health Insurance Portability and Accountability Act of 1996. Pub. L. 104−191 110 Stat. 1936. https://www.congress.gov/bill/104th-congress/house-bill/3103; 1996.

[36] Mukherjee S. Cancer, our genes, and the anxiety of risk-based medicine. Health Aff [Internet] 2018;37(5):817−20. Available from: https://doi.org/10.1377/hlthaff.2018.0344.

[37] Reardon J. The postgenomic condition. University of Chicago Press; 2017.

[38] Grubs RE, Parker LS, Hamilton R. Subtle psychosocial sequelae of genetic test results. Curr Genet Med Rep [Internet] 2014;2(4):242−9. Available from: https:// doi.org/10.1007/s40142-014-0053-7.

[39] Hamilton R. Being young, female, and BRCA positive. Am J Nurs [Internet] 2012;112 (10):26−31. Available from: https://doi.org/10.1097/01.NAJ.0000421021.62295.3b quiz 46, 32.

[40] Supreme Court of California. Tarasoff v. Regents of University of California, 17 Cal.3d 425; 1976.

[41] Supreme Court of Florida. Pate v. Threlkel, 640 So. 2d 183, 186 (Fla. 1st DCA 1994); 1994.

[42] Superior Court of New Jersey. Safer v. Pack, 677 A.2d 1188 (N.J. Super. A.D. 1996); 1996.

[43] Kilbride MK. Genetic privacy, disease prevention, and the principle of rescue. Hastings Cent Rep [Internet] 2018;48(3):10—17. Available from: https://doi.org/10.1002/hast.849.

[44] Liao SM, Mackenzie J. Genetic information, the principle of rescue, and special obligations. Hastings Cent Rep [Internet] 2018;48(3):18—19. Available from: https://doi.org/10.1002/hast.850.

[45] Buchanan A. Ethical responsibilities of patients and clinical geneticists. J Health Care Law Policy [Internet] 1998;1(2):391. Available from: https://digitalcommons.law.umaryland.edu/jhclp/vol1/iss2/6.

[46] Hollands GJ, French DP, Griffin SJ, Prevost AT, Sutton S, King S, et al. The impact of communicating genetic risks of disease on risk-reducing health behaviour: systematic review with meta-analysis. BMJ [Internet] 2016;i1102. Available from: https://doi.org/10.1136/bmj.i1102.

[47] Silarova B, Sharp S, Usher-Smith JA, Lucas J, Payne RA, Shefer G, et al. Effect of communicating phenotypic and genetic risk of coronary heart disease alongside web-based lifestyle advice: the INFORM Randomised Controlled Trial. Heart [Internet] 2019;105 (13):982—9. Available from: https://doi.org/10.1136/heartjnl-2018-314211.

[48] Smerecnik C, Grispen JEJ, Quaak M. Effectiveness of testing for genetic susceptibility to smoking-related diseases on smoking cessation outcomes: a systematic review and meta-analysis. Tob Control [Internet] 2012;21(3):347—54. Available from: https://doi.org/10.1136/tc.2011.042739.

[49] Tan PY, Mitra SR, Amini F. Lifestyle interventions for weight control modified by genetic variation: a review of the evidence. Public Health Genomics [Internet] 2019;21(5—6):1—17. Available from: https://doi.org/10.1159/000499854.

[50] Parker LS. Returning individual research results: what role should people's preferences play? Minn J Law 2012;13(2):449—84.

[51] Appelbaum PS, Roth LH, Lidz CW, Benson P, Winslade W. False hopes and best data: consent to research and the therapeutic misconception. Hastings Cent Rep [Internet] 1987;17(2):20—4. Available from: https://doi.org/10.2307/3562038.

[52] Henderson GE, Churchill LR, Davis AM, Easter MM, Grady C, Joffe S, et al. Clinical trials and medical care: defining the therapeutic misconception. PLoS Med [Internet] 2007;4(11):e324. Available from: https://doi.org/10.1371/journal.pmed.0040324.

[53] Kang JS. Ethical ruminations of a rheumatologist: autoimmunity is an important consideration for immunotherapy trials. Am J Bioeth [Internet] 2018;18(4):75—6. Available from: https://doi.org/10.1080/15265161.2018.1444823.

[54] Appelbaum PS, Parens E, Waldman CR, Klitzman R, Fyer A, Martinez J, et al. Models of consent to return of incidental findings in genomic research. Hastings Cent Rep [Internet] 2014;44(4):22—32. Available from: https://doi.org/10.1002/hast.328.

[55] Manolio TA. Genomewide association studies and assessment of the risk of disease. N Engl J Med [Internet] 2010;363(2):166—76. Available from: https://doi.org/10.1056/NEJMra0905980.

[56] Manolio TA, Collins FS, Cox NJ, Goldstein DB, Hindorff LA, Hunter DJ, et al. Finding the missing heritability of complex diseases. Nat [Internet] 2009;461 (7265):747—53. Available from: https://doi.org/10.1038/nature08494.

[57] Torkamani A, Wineinger NE, Topol EJ. The personal and clinical utility of polygenic risk scores. Nat Rev Genet [Internet] 2018;19(9):581—90. Available from: https://doi.org/10.1038/s41576-018-0018-x.

[58] Regalado A. 23andMe thinks polygenic risk scores are ready for the masses, but experts aren't so sure. Technol Rev [Internet] 2019;. Available from: https://www.technologyreview.com/2019/03/08/136730/23andme-thinks-polygenic-risk-scores-are-ready-for-the-masses-but-experts-arent-so-sure/.

[59] Tandy-Connor S, Guiltinan J, Krempely K, LaDuca H, Reineke P, Gutierrez S, et al. False-positive results released by direct-to-consumer genetic tests highlight the importance of clinical confirmation testing for appropriate patient care. Genet Med [Internet] 2018;20(12):1515—21. Available from: https://doi.org/10.1038/gim.2018.38.

[60] Matthews LJ. Precision (mis)education. Hastings Cent Rep 2020;50(1). Available from: https://doi.org/10.1002/hast.1072 inside front cover.

[61] Spiegelhalter DJ. Understanding uncertainty. Ann Fam Med [Internet] 2008;6 (3):196—7. Available from: https://doi.org/10.1370/afm.848.

[62] Landry LG, Ali N, Williams DR, Rehm HL, Bonham VL. Lack of diversity in genomic databases is a barrier to translating precision medicine research into practice. Health Aff (Millwood) [Internet] 2018;37(5):780—5. Available from: https://doi.org/10.1377/hlthaff.2017.1595.

[63] Mensah GA, Jaquish C, Srinivas P, Papanicolaou GJ, Wei GS, Redmond N, et al. Emerging concepts in precision medicine and cardiovascular diseases in racial and ethnic minority populations. Circ Res [Internet] 2019;125(1):7—13. Available from: https://doi.org/10.1161/CIRCRESAHA.119.314970.

[64] De La Vega FM, Bustamante CD. Polygenic risk scores: a biased prediction? Genome Med [Internet] 2018;10(1):100. Available from: https://doi.org/10.1186/s13073-018-0610-x.

[65] Lachance J, Tishkoff SA. SNP ascertainment bias in population genetic analyses: why it is important, and how to correct it: prospects & overviews. Bioessays [Internet] 2013;35(9):780—6. Available from: https://doi.org/10.1002/bies.201300014.

[66] Borrell LN, Elhawary JR, Fuentes-Afflick E, Witonsky J, Bhakta N, Wu AHB, et al. Race and genetic ancestry in medicine — a time for reckoning with racism. N Engl J Med [Internet] 2021;384(5):474—80. Available from: https://doi.org/10.1056/NEJMms2029562.

[67] All of Us Research Program overview [Internet]. All of Us Research Program | NIH. 2020. Available from: https://allofus.nih.gov/about/program-overview.

[68] Sankar PL, Parker LS. The precision medicine initiative's all of us research program: an agenda for research on its ethical, legal, and social issues. Genet Med [Internet] 2017;19(7):743—50. Available from: https://doi.org/10.1038/gim.2016.183.

[69] Resnicow K, Page SE. Embracing chaos and complexity: a quantum change for public health. Am J Public Health [Internet] 2008;98(8):1382—9. Available from: https://doi.org/10.2105/AJPH.2007.129460.

[70] Taylor-Robinson D, Kee F. Precision public health-the Emperor's new clothes. Int J Epidemiol 2019;48:1—6. Available from: https://www.ncbi.nlm.nih.gov/pmc/articles/PMC6380317/.

[71] Kolarcik CL, Bledsoe MJ, O'Leary TJ. Returning individual research results to vulnerable individuals. Am J Pathol [Internet] 2022;192(9):1218—29. Available from: https://doi.org/10.1016/j.ajpath.2022.06.004.

[72] Botkin JR, Mancher M, Busta ER, Downey AS, editors. National Academies of Sciences, Engineering, and Medicine Committee on the Return of Individual-Specific Research Results Generated in Research Laboratories, Board on Health Sciences Policy, Health and Medicine Division. Returning individual research results to participants: Guidance for a new research paradigm. Washington, D.C.: National Academies Press; 2018.

[73] Rice T. The behavioral economics of health and health care. Annu Rev Public Health [Internet] 2013;34(1):431—47. Available from: https://doi.org/10.1146/annurev-publhealth-031912-114353.

[74] United States Congress. The Genetic Information Nondiscrimination Act of 2008. Pub.L. 110—233, 122 Stat. 881. https://www.congress.gov/bill/110th-congress/house-bill/493; 2008.

[75] Pregelj L, Hwang TJ, Hine DC, Siegel EB, Barnard RT, Darrow JJ, et al. Precision medicines have faster approvals based on fewer and smaller trials than other medicines. Health Aff (Millwood) [Internet] 2018;37(5):724−31. Available from: https://doi.org/10.1377/hlthaff.2017.1580.

[76] Buchanan A, Califano A, Kahn J, McPherson E, Robertson J, Brody B. Pharmacogenetics: ethical issues and policy options. Kennedy Inst Ethics J [Internet] 2002;12(1):1−15. Available from: https://doi.org/10.1353/ken.2002.0001.

[77] Michie M, Cadigan RJ, Henderson G, Beskow LM. Am I a control?: Genotype-driven research recruitment and self-understandings of study participants. Genet Med [Internet] 2012;14(12):983−9. Available from: https://doi.org/10.1038/gim.2012.88.

[78] Beskow LM, Fullerton SM, Namey EE, Nelson DK, Davis AM, Wilfond BS. Recommendations for ethical approaches to genotype-driven research recruitment. Hum Genet [Internet] 2012;131(9):1423−31. Available from: https://doi.org/10.1007/s00439-012-1177-z.

[79] Shirts BH, Parker LS. Changing interpretations, stable genes: responsibilities of patients, professionals, and policy makers in the clinical interpretation of complex genetic information. Genet Med [Internet] 2008;10(11):778−83. Available from: https://doi.org/10.1097/GIM.0b013e31818bb38f.

[80] Wren AA, Park KT. Targeted dosing as a precision health approach to pharmacotherapy in children with inflammatory bowel disease. AMA J Ethics [Internet] 2018;20(9):E841−8. Available from: https://doi.org/10.1001/amajethics.2018.841.

[81] Caulfield T, Zarzeczny A. Defining "medical necessity" in an age of personalised medicine: a view from Canada: insights & perspectives. Bioessays [Internet] 2014;36 (9):813−17. Available from: https://doi.org/10.1002/bies.201400073.

[82] Parker LS, Satkoske VB. Ethical dimensions of disparities in depression research and treatment in the pharmacogenomic era. J Law Med Ethics [Internet] 2012;40 (4):886−903. Available from: https://doi.org/10.1111/j.1748-720x.2012.00718.x.

[83] Parker LS, Brody H. Comparative effectiveness research: a threat to patient autonomy? Health Prog 2011;92(5):64−71.

[84] Joffe S, Lynch HF. Federal right-to-try legislation − threatening the FDA's public health mission. N Engl J Med [Internet] 2018;378(8):695−7. Available from: https://doi.org/10.1056/NEJMp1714054.

[85] Batten J.N. How stratification unites ethical issues in precision health. AMA J Ethics [Internet]. 2018;20(9):E798−E803. https://doi.org/10.1001/amajethics.2018.798.

[86] Tranvåg EJ, Strand R, Ottersen T, Norheim OF. Precision medicine and the principle of equal treatment: a conjoint analysis. BMC Med Ethics [Internet] 2021;22 (1):55. Available from: https://doi.org/10.1186/s12910-021-00625-3.

[87] Aronson SJ, Rehm HL. Building the foundation for genomics in precision medicine. Nat [Internet] 2015;526(7573):336−42. Available from: https://doi.org/10.1038/nature15816.

[88] Richards S, Aziz N, Bale S, Bick D, Das S, Gastier-Foster J, et al. Standards and guidelines for the interpretation of sequence variants: a joint consensus recommendation of the American College of Medical Genetics and Genomics and the Association for Molecular Pathology. Genet Med [Internet] 2015;17(5):405−24. Available from: https://doi.org/10.1038/gim.2015.30.

CHAPTER 8

Precision of diagnostic approaches and individualized therapy toward improving patient outcomes

Loukas G. Chatzis, Ourania Argyropoulou, Konstantinos Panagiotopoulos, Panagiota Palla and Athanasios G. Tzioufas
Department of Pathophysiology, School of Medicine, National and Kapodistrian University of Athens, Athens, Greece

Background

Hippocrates, often dubbed the "Father of Medicine," insightfully asserted that it is far more important to know the person that has a disease than the disease the person has [1]. This simple notion propounded more than 2000 years ago has become the foundation of the practice of precision medicine, being at the core of a solid and effective patient—physician-centered relationship that optimizes health care delivery [2]. When referring to precision medicine, one usually focus on the various targeted treatments, scanning different cells, intracellular molecules, or cytokines that underpin the processes of specific diseases. However, preceding treatment selection, an accurate and individualized diagnostic course of action is essential. High-throughput novel techniques utilizing molecular information (genomic, transcriptomic, proteomic, metabolomic, etc.) have enabled the adoption of tailored approaches running the whole gamut of personal care, from start to finish, and indeed, in the not-too-distant future, from the cradle to the grave. In this chapter, we describe the most recent and important advances in clinical practice related to diagnostic precision medicine approaches across medicine.

Precision medicine is defined as a patient-centered rather than a disease-centered approach to healthcare, where customization of diagnostic, prognostic, or treatment decisions is applied based on individual "high precision" characteristics. These include data from a molecular study of single cells, tissue, or blood, namely transcriptome, protein, and gene expression. The precision medicine paradigm departs from the traditional rather sclerotic notion

The New Era of Precision Medicine
DOI: https://doi.org/10.1016/B978-0-443-13963-5.00006-6
© 2024 Elsevier Inc.
All rights reserved.

and states that "one size does not fit all." Though it was the limitations of the disease-based approach that catalyzed the new paradigm, the latter is earning a growing interest and is already making significant contributions with the emergence of new and promising methods of data analysis in the fields of machine learning and deep learning, combined with advances in biotechnology, such as single-cell RNA sequencing and mass cytometry. We are in a sense putting the cart before the horse when we base our investigations on the clinical phenotype, which may lead to delays in diagnosis or diagnostic errors, instead of focusing on the underlying pathogenetic process that precedes it. Essentially, these modern tools could be used to harness the abundance of evidence that modern medicine bristles with.

Precision medicine in hematology

Among all the various concepts and applications of precision medicine in different aspects of medicine, the field of hematology is a pioneer in the development and implementation of novel techniques and strategies focusing on essential aspects of the practice of medicine, including prevention, diagnostic precision, treatment selection, and optimal treatment [3]. The increasingly more common use of genome sequencing and genomic profiling has been a crucial addition to our arsenal toward a patient- and disease-specific diagnostic and therapeutic plan. Identifying unique gene variants, translocations, copy number changes, and point mutations enables us to directly target them with gene-based therapeutic approaches, such as immunotherapies. Indeed, hematology is brimming with such examples, their wide implementation giving hope to patients struck with a disease that often carries a dismal prognosis. The most striking example is represented by the drug imatinib, a drug that drastically modified the treatment and prognosis of chronic myeloid leukemia (CML) [4]. The turning point was the identification of the BCR::ABL1 fusion gene that produces an abnormal tyrosine kinase that transforms normal blood-forming cells into malignant ones [5]. Imatinib, a tyrosine kinase inhibitor (TKI), blocks this abnormal protein, stalling the division and growth of cancer cells [6]. Imatinib commands a 95% response rate in patients with CML and extends quality-adjusted life by approximately 9 years, while a near-normal life expectancy is achieved by younger patients responding optimally to TKI therapy [7,8]. Nowadays, genetic profiling has been fundamental in the workup of myeloid hematologic malignancies, while its role in lymphoid cancers is still being defined. But even in the field of hematology, and despite the recent advances, the overall impact of precision medicine on the everyday clinical care of patients has been modest, hindered by a multitude of challenges. We are still grappling with questions as to when, to whom, or at what frequency and cost should these diagnostic and therapeutic approaches be employed.

Tools and approaches employed in precision medicine

A number of tools have been developed and applied in the context of precision medicine not only to uncover genetic alterations that drive hematopoietic disorders but also to establish a functional stratification of these mutations, longitudinal "multiple hit" combinations, and the interplay between concurrent mutations [9,10]. For this, a considerable amount of genetic material originating from the neoplastic cell is required. In this regard, hematologic malignancies, especially leukemias, present a fertile ground, given the accessibility of neoplastic cells either by peripheral blood draw or by minimally invasive bone marrow aspirates, during the course of the disease. This enables molecular analyses far beyond the configuration and crude structure of each chromosome afforded with cytogenetics such as karyotyping. Using multicolor fluorescence in situ hybridization (FISH), panels of gene-specific DNA probes annealing to specific target sequence for deletions, aneuploidy, amplification, and gene fusions can help in the diagnosis and provide prognostic information. Recently, with the development of massive parallel sequencing, this being the foundation of next-generation sequencing (NGS), one can search for point mutations, indels, gene fusions, mutation profile/pattern, copy number alterations, and gene expression profiles [2]. Many hot spot panels have been created covering either all coding regions or only the exons of 5−50 genes. These NGS panels do not typically detect chromosomal translocations and other gene fusions and are generally used in conjunction with other methods, such as karyotype and FISH studies. They do, however, possess a very high depth of coverage (often $>1000\times$), which can be extended to include the entire exome (WES), or even the genome (WGS), giving up partly their unique sequence depth ability, since the achievable depth of coverage is determined by the amount of sequencing real estate dedicated to the target of interest. We should bear in mind that we are searching for somatic and not germline mutations. Somatic mutations create a genetic mosaicism that is present only in a small fraction of cells, particularly before their clonal expansion (at the precancerous state), and is further diluted within the tissue, so that their frequency is usually below the detection power of conventional sequencing methods.

Precision diagnostics in myeloid malignancies

As mentioned, CML serves as the best example of the use of genetic profiling to guide diagnostic accuracy and treatment selection. The genetic
aberration (reciprocal translocation) created by the fusion of the bcr and
abl genes on chromosomes 22 and 9, respectively, gives birth to a chimeric BCR-ABL1 gene, located on chromosome 22 (Philadelphia chromosome) [11]. This new fusion gene translates into an oncoprotein, called
BCR-ABL1, characterized by a constitutively activated tyrosine kinase.
From a genetic perspective, CML, a myeloproliferative neoplasm, is a
rather simple disease, serving as a paradigm of how a single genetic mutation can drive carcinogenesis. Identification of the Philadelphia chromosome, the BCR::ABL1 fusion gene, or the BCR::ABL1 fusion mRNA by
conventional cytogenetics, FISH analysis, or reverse transcription polymerase chain reaction (PCR) (qualitative or quantitative RT-PCR),
respectively, suffices to confirm the diagnosis of CML and initiate treatment with a TKI [12].

Polycythemia vera, essential thrombocythemia, and primary myelofibrosis are the remaining myeloproliferative disorders that are negative for
the BCR::ABL1 gene. From a genetic point of view, these diseases are
much more complex and diverse. Despite their genetic and clinical heterogeneity, the determination of driver point mutations in genes, such as
JAK2, calreticulin (CALR), and MPL, is a crucial element of the diagnostic process [13]. Novel targeted treatments have been developed based on
these genetic alterations mainly for the treatment of myelofibrosis [ruxoli-
tinib (Jak1/2 inhibitor)/ fedratinib (Jak2 inhibitor)] [14,15].

For myelodysplastic syndromes (MDS), a group of disorders caused by
bone marrow cells that are poorly formed, immature, and unable to
become healthy blood cells, the genetic landscape is even more complex.
The advent of novel techniques of large-scale and in-depth DNA
sequencing (NGS and WES) has exposed many point mutations that can
aid in diagnosis, patient stratification, prediction of response, and treatment scheme selection. These mutations are not pathogenetic per se but
are partly implicated in the cascade, leading to the abnormal evolutionary
potential of these cells. Their detection requires panel sequencing or analysis of the entire exome or genome. In addition to mutation status (presence or absence), mutational variant allele frequency (VAF) (also known
as mutation allelic burden) is also crucial. VAF is defined as the number of
variant reads divided by the number of total reads, reflecting the

percentage of the clonal burden [16]. A VAF of mutated SF3B1 by more than 10%, along with a cytopenia, irrespective of dysplasia, is considered to define MDS [17]. The same holds true for patients with deletions of the long arm of chromosome 5 [5q], with an additional cytogenetic abnormality [except −7/del[7q]]. Many more molecular variants are included together with clinical and pathologic parameters in individualized prediction models for prognosis generated by artificial intelligence (AI) [18].

Cytogenetics has been used to question the fixed blast percentage as the only distinguishing factor between MDS and acute myeloid leukemia (AML). Genetic mutations are considered MDS-excluding and suggestive of AML, regardless of the blast percentage. These are (1) t(8;21)(q22;q22); RUNX1::RUNX1T1 (previously AML1::ETO); (2) nv(16)(p13.1q22) or t(16;16)(p13.1;q22); CBFB::MYH11; (3) t(15;17)(q22;q21.1); and PML::RARA. Consequently, these patients are eligible for more intense therapies, mirroring their increased responsiveness. Recently, the investigational focus has shifted toward whole transcriptome sequencing (WTS). Based on the transcriptional profile, patients with AML can be reclassified and subdivided into AML refined clusters (e.g., NPM1-mutated AML), carrying a different prognosis or response to treatment. What is more, according to the latest World Health Organization Classification of Hematolymphoid Tumors (WHO5), the fixed blast percentage for defining AML for most of many predefined genetic abnormalities was eliminated [19] (Table 8.1).

Table 8.1: Acute myeloid leukemia with defining genetic abnormalities (no blast % cut-off)

Acute promyelocytic leukemia with PML::RARA fusion (require at least 20% blasts for diagnosis)
Acute myeloid leukemia with RUNX1::RUNX1T1 fusion
Acute myeloid leukemia with CBFB::MYH11 fusion
Acute myeloid leukemia with DEK::NUP214 fusion
Acute myeloid leukemia with RBM15::MRTFA fusion
Acute myeloid leukemia with BCR::ABL1 fusion
Acute myeloid leukemia with KMT2A rearrangement
Acute myeloid leukemia with MECOM rearrangement
Acute myeloid leukemia with NUP98 rearrangement
Acute myeloid leukemia with NPM1 mutation
Acute myeloid leukemia with CEBPA mutation (require at least 20% blasts for diagnosis)

Precision diagnostics in lymphoid malignancies

Although there are unsolved riddles in lymphoid hematologic malignancies aplenty that seek resolution through advanced cytogenetics, their clinical use has been limited and is mainly represented in the face of chronic lymphocytic leukemia (CLL). Being the most common leukemia in adults, it is characterized by a progressive accumulation of functionally incompetent lymphocytes and a rather good prognosis. However, the presence of TP53 mutations (analyzed by Sanger sequencing, PCR, or NGS) and del[17p] and/or del[11q] (analyzed by FISH) are considered both prognostic and predictive markers and are part of the regular pretreatment patient evaluation. Testing for the immunoglobulin heavy variable gene mutation status, evaluated with Sanger sequencing, is also part of the precision medicine diagnostic workup, with patients with an unmutated sequence expressing a more aggressive disease [20]. Similarly, a complex karyotype identified by chromosome-banding analysis [defined by the presence of ≥ 3 chromosomal abnormalities (structural and/or numerical)], a common denominator among many lymphoid malignancies, may be relevant to treatment decision-making. In parallel, mounting evidence suggests the potential prognostic role of many more genomic alterations, such as NOTCH1 mutations, BTK, PLCg2, and BCL2 mutations, and BIRC3 mutation, but their exact prognostic or predictive role remains to be established [21]. The integration of mutation assessment into diagnostic classification and prognostic scoring systems has reformed the way patients with CLL are followed up and treated.

A similar cytogenetic and molecular characterization has been repeatedly performed in acute lymphoblastic leukemia/lymphoma (ALL/LBL), a group of hematologic malignancies of the lymphoid precursor cell. In approximately 80% of patients, conventional karyotype and FISH analysis can reveal a recurring genetic abnormality. The 2016 WHO Classification of ALL/LBL uses cytogenetic and molecular features to subclassify B cell ALL/LBL into nine distinct groups. Higher resolution genomic analysis has revealed many different genetic alterations (deletions and single-nucleotide variants) in ALL/BLL, but a relevant NGS gene panel, as used in AML, is not yet considered common clinical practice.

Lymphomas represent a group of vastly heterogeneous malignancies of the lymphoreticular system. Molecular characterization has been a vital element of the diagnostic workup of lymphomas for decades. Among the complex pathogenetic mechanisms involved, balanced chromosomal translocations form the genetic hallmark. Translocation (14;18)(q32; q21) has been associated with follicular lymphomas, t(11;14)(q13;q32) mark mantle cell lymphomas, and t(2;5)(p23;q35) point to anaplastic lymphomas, to name but a few [22]. Karyotype and FISH analysis have been employed for their detection for years. Additionally, real-time PCR or NGS panels have been used in the last decade to detect point mutations (such as BRAFV600E in hairy cell leukemia and MYD88L265P in Waldenström's macroglobulinemia), revolutionizing the diagnostic approach [23,24]. Seen through this prism, the advent of newer and broader DNA sequencing techniques has provided scientists with tools to decode the genomic intricacies of lymphoid tumors and their clinical sequelae, focusing particularly on aggressive lymphomas. The stratification of aggressive lymphomas, and especially DLBCLs, is lacking clarity and strength and is crying out for a reclassification based on molecular identities. Nowadays, MYC, BCL-2, and BCL-6 translocations carry significant clinical importance in allocating patients into different severity groups (double and triple hit) [25].

Precision medicine in oncology

Early detection and intervention provide the best hope for curative outcomes in cancer, while the most effective strategy is the one that improves the prognosis of patients. However, current methods for noninvasive detection of early-stage tumors have not wholly delivered on the disease's holy grail.

NGS technologies are progressively becoming the platforms of choice to facilitate early diagnosis and targeted therapies and interventions, given their massively parallel sequencing capability, which can be used to simultaneously screen multiple markers, in multiple samples, for a variety of variants (single-nucleotide and multinucleotide variants, insertions and deletions, gene copy number variations, and fusions) [26]. A crucial step in the workflow of targeted NGS is the enrichment of the genomic regions of interest to be sequenced, against the whole genomic background. This ensures that the NGS work is focused on screening predominantly target regions of interest, with minimal off-target sequencing, making it more accurate and economical. PCR-based (PCR- or amplicon-based) and hybridization capture-based methodologies are the two main approaches employed for target enrichment. Next-generation DNA sequencing technology has dramatically advanced clinical oncology through the identification of therapeutic targets and molecular biomarkers, leading to the personalization of cancer treatments, with significantly improved outcomes for many common and rare tumors.

Molecular profiling using tissue NGS has become a standard of care practice, and, recently, circulating tumor DNA (ctDNA) has emerged as a tool for molecular profiling, a predictor of response to systemic treatment, and a powerful way to measure minimal residual disease (MRD) [27].

Liquid biomarkers — circulating tumor cells, cfDNA, and c-miRNA

Liquid biopsy (LB) is the analysis of tumor material for the detection of circulating tumor cells (CTCs) and cell-free circulating nucleic acids from any bodily fluid, including peripheral blood, urine and cerebrospinal fluid, ascites, pleural effusion, etc., and includes a genomic or proteomic assessment. It is a quick and low-budget method with minimal invasiveness, making it readily acceptable to patients, without major side effects [28,29].

Liquid biopsy, analysis of cell-free DNA (cfDNA) in particular, has emerged as a promising noninvasive diagnostic approach in oncology. Abnormal distribution of DNA methylation is one of the hallmarks of many cancers, and methylation changes occur early during carcinogenesis. Systematic analysis of cfDNA methylation profiles is being developed to tackle important clinical facets, including the early detection of cancer, monitoring for MRD, predicting treatment response and prognosis, and tracing the tissue origin [27].

LB currently has two main applications in the clinic: tumor molecular profiling, serving as a complementary approach to tissue molecular profiling, and dynamic characterization of molecular alterations driving acquired resistance. Along with these applications, promising experimental applications of LB include assessment of response dynamics and MRD monitoring. Additionally, compared to tissue biopsies, LB can better capture tumor heterogeneity, as well as clonal selection, under therapeutic pressure.

Compared with conventional tissue biopsy, LB is noninvasive, can be easily repeated, and can provide insight into the tumor burden and the response to treatment. In addition, LB affords a molecular panorama of the primary cancer, avoiding the problem of interference encountered in biopsy results caused by intratumor heterogeneity and sampling bias.

Importantly, LB is a minimally invasive procedure allowing for serial sampling with minimal risk to the patient. Several targeted LB assays have been developed and find application in the clinical routine. (NSCLC, metastatic castration-resistant prostate cancer, ovarian cancer, and breast cancer) [29].

Radiomics

Radiomics is the extraction of quantitative data from medical imaging, that has the potential to specify tumor phenotype. The radiomics approach has the capacity to construct predictive models for treatment response, essential for the application of personalized medicine. The concept underlying radiomics is that medical imaging contains quantitative information that is not discernible by the human eye, but may reflect the underlying pathophysiology of the tissue [30]. In the imaging of cancer, quantitative radiomic features have the potential to characterize tumor phenotype. An important aim of radiomics is to construct predictive models for response to treatment, based on the features of the tumor phenotype derived from medical images. This is essential in the pursuit of precision medicine, in which treatment is tailored based on the characteristics of individual patients and their tumors.

Spatial omics

Spatial omics (SO) technologies combine molecular characterization with spatial resolution. In the context of cancer, SO provide unique insights into tumor molecular architecture, revealing intratumor heterogeneity, and the spatial organization of the tumor immune microenvironment (TME). SO can be subdivided into spatial genomics (SG), spatial chromatin organization (SCO), spatial transcriptomics (ST), and spatial proteomics (SP) [29].

SG assigns DNA sequencing information to spatial location on the cellular and subcellular levels, allowing it to localize specific genomic sequences within a given tumor tissue. SCO provides information on nuclear chromatin conformation, informing on physical proximity between noncontiguous DNA regions, as well as on the interaction between DNA and associated chromatin proteins. ST spatially resolves gene expression within single cells or tissues. As with ST, SP aims to provide a functional image of the tumor by adding spatial information to proteomic profiling on the cellular and subcellular levels.

Precision medicine in autoimmunity

Why autoimmune diseases?

A common feature of autoimmune diseases is the presence of an early stage of active disease, followed by a later stage typified by the damage to the target organ. Current diagnostic approaches often fail at identifying early disease, thus squandering the opportunity for timely intervention that might thwart the progression of the disease. The goal of precision medicine, through molecular stratification of patients, is to predict which subgroups are at risk of disease progression, when might this be expected to occur, and which treatment regimens are most likely to deliver the best results for each patient [31].

With rheumatic diseases, for example, there is a wide use of non-specific therapies, such as corticosteroids and immunosuppressive antimetabolites that, though compatible with an evidence-based approach, seem contradictory to precision medicine principles, by increasing drug-related adverse effects and sacrificing treatment efficacy in particular patient subgroups. Even when more specific therapies are used, for example, targeting cytokines or intracellular kinases, the lack of patient stratification methods hinders individually tailored treatment targets and prompts a "trial and error" framework, to the detriment of quality health care [32].

There is growing evidence that apparently different autoimmune diseases with variable phenotypes can share common pathophysiologic pathways, such as the expression of common clusters of genes or the sharing of peripheral blood cytokine profiles. There is also evidence of precision-based patient grouping predicting the disease's manifestation or the response to treatment. This suggests that the current systems of clinically based classification of disease might be flawed, which opens the way for a shift in our diagnostic approaches, as well as making a new drug target selection the focus of research [33].

Why not yet?

Systemic autoimmune diseases are heterogeneous entities, relatively rare, and typically evolve from complex interactions between a genetic background facilitating immunologic autoreactivity and various incompletely identified environmental factors, often years before the clinical manifestations of the disease emerge [34].

First, the heterogeneity of clinical phenotypes has led to the creation of composite indexes of global disease activity for research purposes, used as end points in clinical trials. A common scenario observed is the diversity in the treatment response, with some subgroups of patients responding poorly, while others showing complete remission of disease. However, given the fact that in the same patient, different organ manifestations might arise through different pathogenetic pathways, the lack of organ-specific composite indexes has stifled our ability to reliably assess the effects of treatment, even with a patient customized along a subpopulation stratification [35].

Second, with continuous upgrades in the standard of care, it is increasingly more difficult to show statistically significant superiority in responses between different treatment regimens, leading to growing calls for an increase in the population size under study. However, precision medicine techniques tend to be expensive, and a great amount of funding is needed to achieve statistically significant results translatable into modifications in clinical practice.

Third, the time interval between the onset of subclinical disease and its clinical manifestations means that, to obtain evidence regarding the pathogenesis of early disease, asymptomatic individuals must be "blindly" screened for years. Again, the use of precision medicine in these settings is financially onerous and time-consuming and not widely applied [36].

Illustrative examples

Type I interferon signature gene (ISG) expression is the archetype of precision medicine use in autoimmunity. Several autoimmune diseases, such as systemic lupus erythematosus (SLE), dermatomyositis, Sjogren syndrome, scleroderma, and mixed connective disease, have shown increased ISG expression in variable frequency. Anifrolumab, a monoclonal antibody against the type I interferon receptor, has been shown in clinical trials to reduce cutaneous SLE disease activity and SLE flares and allow a reduction in glucocorticoid dosage. Interestingly, these effects are observed only in patients with high ISG expression at diagnosis, while subjects with normal ISG expression showed no treatment benefit. Sifalimumab, a monoclonal antibody against IFN-α, has similarly shown efficacy in patients with SLE and high baseline ISG expression, with ISG suppression being more robust in SLE subgroups with lower disease activity, implying that to achieve ISG suppression in states of high disease activity, a higher drug dosage is needed [37].

Rheumatoid arthritis (RA) is a systemic autoimmune disease characterized by an inflammatory erosive arthritis and frequently by a B-cell-mediated autoantibody production, such as rheumatoid factor (RF) and anticitrullinated protein antibodies (ACPA). Rituximab is a monoclonal antibody directed against CD20, a surface marker expressed mainly on B cells. Clinical trials have shown better treatment response in subgroups positive for RF and ACPA autoantibodies, another example of biomarker-guided therapeutic decisions.

Not all precision medicine techniques need to be costly. In IgG4-related disease, a relatively rare disease in the practice of most clinicians, an AI machine learning model has been developed, using solely a patient's clinical and laboratory characteristics, most of which are regularly measured in clinical practice. This model differentiates between IgG4-RD and its mimics with relatively high sensitivity and specificity and can aid in difficult diagnostic dilemmas [38].

Extrapolating from prognostic models developed for hematologic malignancies, the neutrophil to lymphocyte ratio (NLR) has been suggested, though not sufficiently studied, as a biomarker to predict the response to treatment in SLE. Specifically, cyclophosphamide, an alkylating agent targeting B and T lymphocytes, is used with severe SLE manifestations, such as lupus nephritis. Since a low NLR ratio, arising from disproportionately high lymphocytes, is associated with high BCL2 gene expression, an antiapoptosis molecule favoring autoreactive lymphocyte survival, it is being considered as a potential predictor of response to cyclophosphamide in SLE patients.

Newer technologies have begun taking center stage over the research field in hope of identifying novel drug targets and developing a deeper understanding of the complex pathogenesis of autoimmunity. In RA, single-cell analysis using RNA sequencing and mass cytometry has been used in synovial tissue samples to better characterize gene and protein expression. Testing samples from multiple time points along the disease's course allows for the identification of specific expansion of a lymphocyte population and its evolution in different stages of the disease [39]. Illustrating exactly which proteins are overexpressed in the main effector cells and the timeframe to target them forges the base for the development of the optimal treatment regimen, characterized both by accuracy in the drug target and by precision in the timing of an intervention to avert progression of the disease.

In giant cell arteritis (GCA), a systemic vasculitis characterized by frequent relapses, nuclear magnetic resonance (NMR) metabolomics has recently been used to measure disease activity. The foundation of GCA therapy is the induction of remission by high dose of glucocorticoids, occasionally combined with another agent, followed by glucocorticoid dose tapering. Inflammation metabolites, detected and quantified by NMR spectrometers in serum samples, serve to discriminate between disease activity and remission, potentially recognizing impending relapse and guiding glucocorticoid tapering, thus forestalling both prolonged and unnecessary glucocorticoid exposure with its adverse effects, but also tempering overly eager glucocorticoid tapering that might result in disease relapse and thus an overall increase in glucocorticoid use to retain low disease activity.

Pemphigus vulgaris is a rare chronic, mucosal blistering disease characterized by autoantibodies against desmoglein 3. Chimeric autoantibody receptor (CAAR) T cells have been developed that express a recombinant receptor with the ability to target antidesmoglein-3 autoantibody-producing B cells, in mouse models. Reproducing these experiments in humans will allow the broadening of the available treatment options against the scourge of autoimmunity.

Overall, precision medicine is a promising field of research in systemic rheumatic diseases. Though its implementation in clinical practice has not yet begun, there are valid theoretical reasons supported by existing experimental data and small observational studies that merit our attention, our time, and resources toward the building of a patient-tailored treatment approach. Whether it is the question of prognosis and risk stratification, the issue of early diagnosis and prevention of the disease-associated injury, or biomarker-guided therapeutic decisions, precision medicine creates the framework that would allow us to imagine the heretofore unimagined possibility of looking before and beyond what is clinically apparent, while all along practicing the principle and breathing life to the principle that "One size does not fit all."

Precision medicine in infectious diseases

The successful treatment of infectious diseases begins with the rapid and accurate diagnosis of the pathogenic organisms. For a long time, traditional culture methods have been the gold standard for pathogen detection, though their time-consuming nature and low sensitivity fail the clinical requirements of fast and accurate diagnosis [40]. The recent invention of smear fluorescent detection and ELISA seem to meet the requirement of rapid diagnosis and are already in wide use. However, the sensitivity of these methods varies with the infectious agent (physical access of the fluorescent probe to the infectious agent is not always guaranteed) and the stage of infection (antigen or antibody response may be missed or entirely absent in the sample taken). To increase the sensitivity of pathogen detection, several nucleic acid-based molecular methods have been developed in the past few years, including conventional PCR, nested PCR, real-time PCR, and loop-mediated isothermal amplification (LAMP) and sequencing [41]. Nevertheless, the sensitivity, accuracy, and repeatability of these technologies still do not satisfy the requirements of clinical practice, especially in samples with a low yield of pathogenic DNA, and there remains an urgent need for new methods for quick and precision diagnosis of microbial infections. To these, it can be added the diagnostic challenge of identifying multiple pathogens with a single test. In the past two decades, significant advances in clinical microbiology practice have been made, leading to the development of novel molecular diagnostic platforms.

In the following section, we describe some widely used diagnostic methods.

Multiplex PCR (mPCR) tests (also known as "syndromic" panels) combine tests for numerous pathogens and resistance genes into a single test and have changed the way we diagnose infections, leading to improved patient care and clinical workflow. These syndromic panels can impact infection control, antimicrobial stewardship, and patient outcomes by significantly reducing the time to diagnosis and clinical decision making [42]. Syndromic diagnostic panels are now commercially available in the diagnosis of serious infections that hit the bloodstream and the respiratory, gastrointestinal (GI), and central nervous systems [43].

Mass spectrometry

The use of mass spectrometry (MS) with microbial infectious diseases began in the 1970s, when gas chromatography–MS (GC–MS) was introduced to identify microorganisms. In daily practice, clinical specimens, such as sputum, pus, swab sticks, and urine, are first developed on medium, and microorganisms are isolated as colonies. The pathogens are evaluated based on observations of the cultured colonies, the patient's clinical information, microscopy using Gram-staining, and the number of microorganisms. The ID of the microorganisms is then obtained using a MALDI-TOF MS procedure. The process can be summarized as follows:

A colony is transferred to the sample plate, which will later be exposed to laser irradiation. When the matrix has been added and dried, the plate is inserted into the MS and then irradiated with the laser, causing the proteins in the bacteria (primarily ribosomal proteins) to be ionized; the time required for these ionized proteins to fly to the detector determines the mass-to-charge ratio (m/z) of the component proteins, and the intensity of the signal provides the mass spectrum (pattern). The mass spectrum thus obtained is then compared to those in a reference database to identify the bacteria.

Metagenomic next-generation sequencing

Clinical metagenomic NGS (mNGS) is a new promising method of detecting DNA or RNA originating from infectious organisms (viral, bacteria, fungal, and parasitic) with a higher sensitivity and a faster turnaround than conventional methods. It can practically sequence all nucleic acids from a sample (such as serum, CSF, and bronchoalveolar lavage fluid), select those not belonging to the human host background cell-free DNA (cfDNA), and assign those to their reference genomes. It can be used either as a targeted or as a hypothesis-free, shotgun sequencing untargeted approach. MNGS is becoming more widely available and is employed for pathogen detection in diverse sample sizes for various indications. The Karius and DISQVER tests are two commercially available detection tests in the USA and Europe, respectively [44].

Digital droplet polymerase chain reaction

Digital droplet PCR (ddPCR) has emerged in recent years as a promising tool for pathogen detection. This method has been used to identify patients with early stages of cancer, to profile environmental microorganisms, and to detect human pathogens in the food chain. Application of ddPCR has also been expanded for the identification of pathogens in infectious diseases [40].

The ddPCR principle was first conceived in the 1990s, with the aim of quantifying nucleic acid molecules at the single molecule level. Unlike real-time PCR, ddPCR technology can quantify nucleic acids without a standard curve. The key principle of ddPCR is to distribute a nucleotide-containing sample to thousands of independent partitions. There are several ways to create these microdroplets: manual partitioning, immiscible liquid chemistry, atomization, and so on. The ideal droplets contain only a single-target DNA molecule—or none. The microfluidics are generated in a droplet generator, and each droplet is then read one at a time by a specialized droplet reader. To determine targeted DNA copies without bias, targeted DNA templates must be randomly distributed and micro-droplets should be produced in great numbers. These partitions can then be amplified individually through thermal cycling. Finally, positive partitions with the target sequence and negative partitions without the target sequence are checked and calculated using Poisson's law of small numbers. With these advantages, ddPCR has now been used to detect pathogens, gene mutations, gene copies, and DNA modifications. Importantly, ddPCR technology has an advantage in detecting pathogenic microorganisms in samples with low levels of pathogenic DNA.

While central clinical laboratories offer sensitive and specific assays, such as blood culture, high-throughput immunoassays, PCR, and MS tests, they are often time and labor intensive, costly, and dependent on sophisticated instruments and well-trained operators. On the other hand, point-of-care (POC) tests provide rapid "on-site" results and, in resource-limited settings, contribute to timely and proper treatment.

Point-of-care tests (POCTs), such as compact molecular diagnostic systems, lateral flow assays, microfluidics, plasmonic technologies, and paper-based assays, detect a variety of infectious diseases-related biomarkers, including virus particles, nucleic acids, proteins, and antibodies. They serve as the foundation of "patient centralized" diagnosis and treatment of infectious diseases, marching toward inexpensive, robust, and portable solutions [45].

Precision medicine in gastroenterology

Precision medicine has the potential to revolutionize the field of gastroenterology by tailoring diagnosis and treatment to individual patients, based on their unique genetic, microbiome, and clinical characteristics. There are several GI diseases for which a personalized approach could be implemented, including inflammatory bowel disease (IBD), nonalcoholic fatty liver disease (NAFLD) [46], and autoimmune liver disease (AILD) [47].

Over the past few decades, there have been significant advances in both the understanding and the therapeutic management of IBD, resulting in a range of new therapeutic agents that target specific components of gut inflammation, such as cytokines, receptors, adhesion molecules, and signaling pathways. Despite these advances, the efficacy of these treatments remains unpredictable, with only a subset of patients experiencing a therapeutic benefit, and even that may not be sustained over time. One important reason for this is that current IBD drugs are either nonspecific antiinflammatory agents or biologics that target single components of complex biological processes. To overcome this limitation, we need agents that specifically target the key controllers underpinning the biological complexity of IBD, and this requires the adoption of advanced bioinformatics tools, such as systems biology, that can integrate all components of the disease process into a network medicine [48]. Network medicine is a comprehensive approach that encompasses all the components (omes) of IBD and utilizes computational tools to integrate them into a network called the IBD interactome. By analyzing the IBD interactome, the essential molecular drivers of gut inflammation can be identified and targeted by specific drugs, thereby increasing the likelihood of therapeutic success. The identification of patient subtypes with IBD through computational methods is crucial in developing precision medicine therapeutics that have better efficacy and fewer side effects. In the next decade or so, IBD therapy is expected to shift from a physician-based approach that uses broad antiinflammatory agents to treat phenotypically similar but, crucially, biologically heterogeneous patients, to an omics-based network medicine approach that treats molecularly homogeneous patients with highly specific and customized, personalized medicine drugs [49,50].

In the context of NAFLD, precision medicine can be used to identify the underlying factors that contribute to the development of the disease in a particular patient and to develop targeted therapies that address those factors. The term NAFLD encompasses various disease subphenotypes, each with potentially different natural history, prognosis, and response to therapy. Proper evaluation of the factors contributing to the pathogenesis and clinical expression of NAFLD could help tailor the management of the disease. The diagnostic process should go beyond standard assessments using biomarkers and imaging techniques. The "LDE system" provides a synthetic descriptor of variables related to the severity of liver pathology (L), individual patient factors (D), and extrahepatic involvement (E). Noninvasive diagnostic tools are available to estimate the severity of liver fibrosis in NAFLD, and recent studies suggest modifying thresholds according to a patient's ethnicity. Disease determinants, such as sex, genetics, microbiota, endocrine and metabolic assessments, and physical activity, should also be considered in individual patients. Extrahepatic involvement can include metabolic derangements, cardiovascular disease, and nonhepatic cancer. Categorizing patients can improve on risk stratification and follow-up protocols. The risk of developing NAFLD and related complications has been linked to genetic factors, and genome-wide association studies have identified several genes including *PNPLA3*, *TM6SF2*, *MBOAT7*, *GCKR*, and *HSD17B13*, associated with the development of NAFLD and its severity. The *PNLA3 I148* variant has been extensively characterized and carries an increased risk of developing NAFLD and a higher risk of NASH and HCC per allele. The use of polygenic risk scores (PRS) is being investigated to identify patients at risk of developing more severe disease or complications such as HCC [9]. The primary treatment for ALD and NAFLD is to focus on reducing alcohol consumption and overnutrition. A better understanding of the gut-liver connection is needed to prevent or even reverse ALD and NAFLD by targeting the gut microbiome. In the future, diagnosing patients by characterizing their microbiome and metabolome could become part of regular practice to tailor approaches of microbiome treatment. Rather than using therapies that modify the entire gut microbiome, precision microbiome therapies will be the norm, using engineered bacterial strains or drugs that target specific bacterial enzymes. Future treatments may even include "biopower plants" made of engineered or naturally occurring gut bacteria that produce antiinflammatory peptides or antioxidants [10]. For example, introducing urease-producing bacteria to prevent hepatic encephalopathy or targeting C. difficile's virulence-conferring factors are potential applications of precision microbiome therapies.

AI has the potential to revolutionize the field of AILD by facilitating earlier diagnosis, enabling personalized treatment, and achieving improved patient outcomes. Digital pathology involves transforming analog data in histopathology slides into digital file formats, while computational pathology is the process of extracting and handling data from digital slides using machine learning algorithms. Whole slide imaging (WSI) technology coupled with deep learning (DL) can improve the efficiency and accuracy of the diagnostic process and offer quantitative prognostic information to support choice of therapies. Liver biopsy is an essential component of the diagnostic process for AILDs, providing valuable information for treatment and prognostication. The use of AI-assisted computational pathology may aid in discovering pathogenetic clues, identifying core histological features, and standardizing the list of differential diagnosis. The role of imaging in primary sclerosing cholangitis (PSC) and autoimmune hepatitis (AIH) could reform the diagnosis and management of AILD. While MRCP is the main noninvasive imaging method for PSC diagnosis, there is still a lack of radiological features specific to PSC that can reliably exclude other causes of cholangiopathy. AI and radiomics have the potential to improve diagnostic accuracy, risk stratification, and prediction models for PSC and AIH. For example, MRCP + software can provide quantitative metrics of the biliary tree, generating data on biliary tree volume, median diameter, and strictures and dilatations. Additionally, multiparametric MRI (mpMRI) using iron-corrected T1 (cT1) relaxation maps can provide an accurate noninvasive quantitative biomarker of liver fibrosis and inflammation, with higher cT1 values correlating with a higher risk of loss of biochemical remission in AIH. Finally, ML has potential application on large databases created by genome-wide association studies (GWAS) [16]. Supervised learning approaches can be used to classify cases and controls based on SNPs and identify novel predictive features associated with the phenotype. Unsupervised approaches can be used to cluster patients according to the genotype data and investigate whether these novel groups have different clinical presentations, trajectories, and indeed treatment responses.

Precision medicine has the distinct potential to improve patient outcomes in gastroenterology by providing tailored treatments; by identifying individuals at high risk for developing a disease that might allow for earlier intervention and potentially prevention; by identifying subgroups of patients who are more likely to respond to certain therapies, thus ensuring more efficient and effective clinical trials; and indeed, by identifying more efficient and cost-effective treatments. Precision medicine does, however, face its own challenges related to limited data available, cost, privacy concerns, and ethical considerations. Careful consideration of the pros and cons is needed and necessary to ensure that precision medicine is used in an effective manner and lives up to its considerable potential.

References

[1] Konstantinidou MK, Karaglani M, Panagopoulou M, Fiska A, Chatzaki E. Are the origins of precision medicine found in the corpus hippocraticum? Mol Diag Ther 2017;21(6):601−6.
[2] Valent P, Orfao A, Kubicek S, Staber P, Haferlach T, Deininger M, et al. Precision medicine in hematology 2021: definitions, tools, perspectives, and open questions. HemaSphere 2021;5(3):e536.

[3] Wästerlid T, Cavelier L, Haferlach C, Konopleva M, Fröhling S, Östling P, et al. Application of precision medicine in clinical routine in haematology—challenges and opportunities. J Intern Med 2022;292(2):243—61.

[4] Hochhaus A, Larson RA, Guilhot F, Radich JP, Branford S, Hughes TP, et al. Long-term outcomes of imatinib treatment for chronic myeloid leukemia. N Engl J Med 2017;376(10):917—27.

[5] Rowley JD. Letter: A new consistent chromosomal abnormality in chronic myelogenous leukaemia identified by quinacrine fluorescence and Giemsa staining. Nature. 1973;243(5405):290—3.

[6] Druker BJ, Talpaz M, Resta DJ, Peng B, Buchdunger E, Ford JM, et al. Efficacy and safety of a specific inhibitor of the BCR-ABL tyrosine kinase in chronic myeloid leukemia. N Engl J Med 2001;344(14):1031—7.

[7] Jamy O, Godby R, Sarmad R, Costa LJ. Survival of chronic myeloid leukemia patients in comparison to the general population in the tyrosine kinase inhibitors era: a US population-based study. Am J Hematol 2021;96(7) E265-e8.

[8] Radivoyevitch T, Weaver D, Hobbs B, Maciejewski JP, Hehlmann R, Jiang Q, et al. Do persons with chronic myeloid leukaemia have normal or near normal survival? Leukemia. 2020;34(2):333—5.

[9] Upadhyay Banskota S, Khanal N, Marar RI, Dhakal P, Bhatt VR. Precision medicine in myeloid malignancies: hype or hope? Curr Hematol Malignancy Rep 2022;17(6):217—27.

[10] Ley TJ, Miller C, Ding L, Raphael BJ, Mungall AJ, Robertson A, et al. Genomic and epigenomic landscapes of adult de novo acute myeloid leukemia. N Engl J Med 2013;368(22):2059—74.

[11] Kang ZJ, Liu YF, Xu LZ, Long ZJ, Huang D, Yang Y, et al. The Philadelphia chromosome in leukemogenesis. Chin J Cancer 2016;35:48.

[12] Jabbour E, Kantarjian H. Chronic myeloid leukemia: 2020 update on diagnosis, therapy and monitoring. Am J Hematol 2020;95(6):691—709.

[13] Arber DA, Orazi A, Hasserjian RP, Borowitz MJ, Calvo KR, Kvasnicka HM, et al. International Consensus Classification of Myeloid Neoplasms and Acute Leukemias: integrating morphologic, clinical, and genomic data. Blood 2022;140(11):1200—28.

[14] Verstovsek S, Mesa RA, Gotlib J, Levy RS, Gupta V, DiPersio JF, et al. A double-blind, placebo-controlled trial of ruxolitinib for myelofibrosis. N Engl J Med 2012;366(9):799—807.

[15] Talpaz M, Kiladjian JJ. Fedratinib, a newly approved treatment for patients with myeloproliferative neoplasm-associated myelofibrosis. Leukemia 2021;35(1):1—17.

[16] Deng J, Wu X, Ling Y, Liu X, Zheng X, Ye W, et al. The prognostic impact of variant allele frequency (VAF) in TP53 mutant patients with MDS: a systematic review and meta-analysis. Eur J Haematol 2020;105(5):524—39.

[17] Malcovati L, Karimi M, Papaemmanuil E, Ambaglio I, Jädersten M, Jansson M, et al. SF3B1 mutation identifies a distinct subset of myelodysplastic syndrome with ring sideroblasts. Blood 2015;126(2):233—41.

[18] Döhner H, Wei AH, Löwenberg B. Towards precision medicine for AML. Nat Rev Clin Oncol 2021;18(9):577—90.

[19] Khoury JD, Solary E, Abla O, Akkari Y, Alaggio R, Apperley JF, et al. The 5th edition of the World Health Organization Classification of Haematolymphoid Tumours: Myeloid and Histiocytic/Dendritic Neoplasms. Leukemia 2022;36 (7):1703—19.

[20] Shadman M. Diagnosis and treatment of chronic lymphocytic leukemia: a review. JAMA 2023;329(11):918—32.

[21] Almasri M, Amer M, Ghanej J, Mahmoud AM, Gaidano G, Moia R. Druggable molecular pathways in chronic lymphocytic leukemia. Life (Basel, Switz) 2022;12(2).

[22] Blombery PA, Wall M, Seymour JF. The molecular pathogenesis of B-cell non-Hodgkin lymphoma. Eur J Haematol 2015;95(4):280−93.

[23] Tiacci E, Trifonov V, Schiavoni G, Holmes A, Kern W, Martelli MP, et al. BRAF mutations in hairy-cell leukemia. N Engl J Med 2011;364(24):2305−15.

[24] Treon SP, Xu L, Yang G, Zhou Y, Liu X, Cao Y, et al. MYD88 L265P somatic mutation in Waldenström's macroglobulinemia. N Engl J Med 2012;367(9):826−33.

[25] Alizadeh AA, Eisen MB, Davis RE, Ma C, Lossos IS, Rosenwald A, et al. Distinct types of diffuse large B-cell lymphoma identified by gene expression profiling. Nature 2000;403(6769):503−11.

[26] Singh RR. Target enrichment approaches for next-generation sequencing applications in oncology. Diagnostics (Basel, Switz) 2022;12(7).

[27] Luo H, Wei W, Ye Z, Zheng J, Xu RH. Liquid biopsy of methylation biomarkers in cell-free DNA. Trends Mol Med 2021;27(5):482−500.

[28] Zygulska AL, Pierzchalski P. Novel diagnostic biomarkers in colorectal cancer. Int J Mol Sci 2022;23(2).

[29] Akhoundova D, Rubin MA. Clinical application of advanced multi-omics tumor profiling: shaping precision oncology of the future. Cancer Cell 2022;40(9):920−38.

[30] Chetan MR, Gleeson FV. Radiomics in predicting treatment response in non-small-cell lung cancer: current status, challenges and future perspectives. Eur Radiol 2021;31(2):1049−58.

[31] Toro-Domínguez D, Alarcón-Riquelme ME. Precision medicine in autoimmune diseases: fact or fiction. Rheumatol (Oxford, Engl) 2021;60(9):3977−85.

[32] Kissler S. Toward a precision medicine approach for autoimmunity. Proc Natl Acad Sci USA 2022;119(19) e2204841119.

[33] Desvaux E, Aussy A, Hubert S, Keime-Guibert F, Blesius A, Soret P, et al. Model-based computational precision medicine to develop combination therapies for autoimmune diseases. Expert Rev Clin Immunol 2022;18(1):47−56.

[34] Chatzis L, Vlachoyiannopoulos PG, Tzioufas AG, Goules AV. New frontiers in precision medicine for Sjogren's syndrome. Expert Rev Clin Immunol 2021;17(2):127−41.

[35] Chatenoud L. Precision medicine for autoimmune disease. Nat Biotechnol 2016;34 (9):930−2.

[36] Wampler Muskardin TL, Paredes JL, Appenzeller S, Niewold TB. Lessons from precision medicine in rheumatology. Multiple Scler (Houndmills, Basingstoke, Engl) 2020;26(5):533−9.

[37] De Ceuninck F, Duguet F, Aussy A, Laigle L, Moingeon P. IFN-α: A key therapeutic target for multiple autoimmune rheumatic diseases. Drug Discov Today 2021;26 (10):2465−73.

[38] Yamamoto M, Nojima M, Kamekura R, Kuribara-Souta A, Uehara M, Yamazaki H, et al. The differential diagnosis of IgG4-related disease based on machine learning. Arthritis Res Ther 2022;24(1):71.

[39] Zhang F, Wei K, Slowikowski K, Fonseka CY, Rao DA, Kelly S, et al. Defining inflammatory cell states in rheumatoid arthritis joint synovial tissues by integrating single-cell transcriptomics and mass cytometry. Nat Immunol 2019;20(7):928−42.

[40] Chen B, Jiang Y, Cao X, Liu C, Zhang N, Shi D. Droplet digital PCR as an emerging tool in detecting pathogens nucleic acids in infectious diseases. Clin Chim Acta 2021;517:156−61.

[41] Fakruddin M, Mannan KS, Chowdhury A, Mazumdar RM, Hossain MN, Islam S, et al. Nucleic acid amplification: alternative methods of polymerase chain reaction. J Pharm Bioallied Sci 2013;5(4):245−52.

[42] Dumkow LE, Worden LJ, Rao SN. Syndromic diagnostic testing: a new way to approach patient care in the treatment of infectious diseases. J Antimicrob Chemother 2021;76(Suppl. 3):iii4−iii11.

[43] Kim H, Huh HJ, Park E, Chung DR, Kang M. Multiplex molecular point-of-care test for syndromic infectious diseases. Biochip J 2021;15(1):14–22.

[44] Tsuchida S, Umemura H, Nakayama T. Current status of matrix-assisted laser desorption/ionization-time-of-flight mass spectrometry (MALDI-TOF MS) in clinical diagnostic microbiology. Molecules (Basel, Switz) 2020;25(20).

[45] Chen H, Liu K, Li Z, Wang P. Point of care testing for infectious diseases. Clin Chim Acta 2019;493:138–47.

[46] Bluemel S, Williams B, Knight R, Schnabl B. Precision medicine in alcoholic and nonalcoholic fatty liver disease via modulating the gut microbiota. Am J Physiol Gastrointest liver Physiol 2016;311(6) G1018-g36.

[47] Fiocchi C, Dragoni G, Iliopoulos D, Katsanos K, Ramirez VH, Suzuki K. Results of the seventh scientific workshop of ECCO: precision medicine in IBD—what, why, and how. J Crohn's Colitis 2021;15(9):1410–30.

[48] Menche J, Sharma A, Kitsak M, Ghiassian SD, Vidal M, Loscalzo J, et al. Disease networks. Uncovering disease-disease relationships through the incomplete interactome. Science (New York, NY) 2015;347(6224):1257601.

[49] de Souza HSP, Fiocchi C, Iliopoulos D. The IBD interactome: an integrated view of aetiology, pathogenesis and therapy. Nat Rev Gastroenterol Hepatol 2017;14(12):739–49.

[50] Yadav A, Vidal M, Luck K. Precision medicine – networks to the rescue. Curr Opin Biotechnol 2020;63:177–89.

CHAPTER 9

"Multiomics in precision medicine"

Konstantinos Katsos[1,2], Ashis Dhar[1,2] and F.M. Moinuddin[1,2]
[1]Mayo Clinic Neuro-Informatics Laboratory, Mayo Clinic, Rochester, MN, United States
[2]Department of Neurologic Surgery, Mayo Clinic, Rochester, MN, United States

Introduction

Precision medicine attempts to determine accurate and sensitive indicators of health status, identify disease in the early stages of development, revert or diminish disease progression, and improve outcomes by tailoring treatment to the individual's needs. This effort requires a comprehensive understanding of individual disease processes and the underlying pathophysiological intricacies. Biological functions can be conceptualized as the summation of interacting molecules participating in physical and biochemical processes across multiple levels [1]. Technological advancements have facilitated a more holistic study of such levels, moving away from the intrinsic mechanisms and focusing on their overall interactions and their effect on the system [2]. The addition of the suffix "-omics" to a molecular term denotes the collective characterization and quantification of the respective biomolecules within that level, which participate in structural and functional biological processes [3,4]. For example, genetics studies the composition and function of individual genes, whereas genomics examines the genome as a whole and scrutinizes the overall effect of the genome on the biological system [4]. A comprehensive understanding of these highly complicated and intricate processes necessitates interpreting and integrating information across multiple levels, giving rise to the study of "multiomics" [4,5]. Although this approach holds excellent potential in understanding the genotype-to-phenotype relationship and delivering personalized treatment, it inherently poses a "big data" problem [5]. Consequently, a robust framework is crucial for storing, handling, and processing data. In this chapter, we will identify some of the critical elements of "multiomics," provide examples of current applications and the role of "multiomics" in precision medicine, and outline future perspectives and challenges.

The New Era of Precision Medicine
DOI: https://doi.org/10.1016/B978-0-443-13963-5.00011-X
© 2024 Elsevier Inc.
All rights reserved.

Different types of "omic" data

As per the central dogma theory, genetic information flows from DNA to RNA to proteins, and each of these steps can be further examined independently, giving rise to different types of "omics" [4,6]. We discuss the most widely used types of "omics" as follows:

Genomics studies the cell's genetic blueprint and investigates the presence or absence of genes and focuses on their association with disease, response to treatment, or prognosis [4]. Genome-wide association studies (GWAS) have identified thousands of genetic variants associated with complex diseases across various populations [7]. In GWAS, the individual's genotype is screened for more than a million genetic markers, and statistically significant differences in alleles between cases and controls are considered evidence of association. Genomics utilizes genotype arrays and next-generation sequencing (NGS) for whole-genome sequencing, exome sequencing, and targeted sequencing to extrapolate data and identify patterns. In fact, most early efforts in precision medicine have utilized genomic information to aid diagnosis, stratify risk factors, and predict response to treatment [8,9]. Given that specific disease-causing sequences and their pathophysiological mechanisms have yet to be established, clinical applications of genomics have been limited. However, despite the genetic heterogeneity of cancer, both between tumor types and the same types of tumor but different individuals, the field of genomics in oncology has made significant progress [10,11]. Identification of chromosomal abnormalities and copy variations, such as the BCR-ABL fusion genes, has provided insight into abnormal cell growth and treatment resistance in cancer [12]. In addition, further advancements in sequencing techniques and emerging data have revealed that genetic profiles in cancer are even more complicated than initially thought, particularly when attempting to distinguish "driver" mutations from phenotypically neutral "passenger" mutations. The Cancer Genome Atlas (TCGA) has been established as a response to unravel the mutational burden in tumors and characterize the underlying pathophysiologic mechanisms [13–17]. Nonetheless, it is evident that genomic characterization of an individual or genomic profiling of tumor cells does not sufficiently predict the risk of developing cancer, nor response to treatment or risk of recurrence.

Transcriptomics studies the transcribed genetic material by quantitatively and qualitatively analyzing RNA levels genome wide [18]. Large transcriptomic studies have shown that noncoding RNA elements play essential roles in physiological and pathological processes. Noncoding RNAs and their dysregulation is being intensively studied, as they may embody potential biomarkers or therapeutic targets. Probe-based arrays, RNA sequencing, and targeted sequencing are often employed to conduct such analyses [5]. Of particular interest is the expression quantitative trait loci (QTL) analysis, which examines the relationship between DNA sequence variations and changes in gene expression [19]. Such analyses hold the potential to reveal functional mechanisms by which genetic variation leads to disease. Multiple initiatives, including the Gene Expression Omnibus (GEO), the TCGA, and the Stand Up To Cancer-Prostate Center Foundation Project, have used transcriptomics to characterize gene expression changes and their associated effects on tumor formation and growth [20,21]. Transcriptomic research has also established gene expression signatures, which predict patient outcome, treatment response, prognosis, and recurrence risk in a variety of cancer types [22−24]. In addition, single-cell gene expression analysis has been carried out to study cell heterogeneity in cancer [25,26].

Epigenomics examines genome-wide, reversible modifications of DNA and/or associated proteins that participate in transcription regulation and ultimately determine cell fate [27]. Methylation and histone acetylation are well-known examples of such epigenetic control. Epigenome-wide association studies (EWAS) have located methylated regions of DNA, which can act as indicators of disease status in multiple pathologic states [28]. EWAS identify DNA modifications by performing NGS, eQTL analyses, and single-cell gene expression analysis to identify the significance of cell-specific somatic mutations on gene expression within individual cells. Characteristically, MGMT promoter methylation has been broadly utilized as a predictor of treatment response to temozolomide in glioblastoma [29]. Similarly, epigenetic control at the miRNA level has proved to be associated with doxorubicin resistance in lung cancer [30]. miRNA and lncRNA may provide promising therapeutic targets for novel therapeutic strategies. However, despite the promising potential, it is unlikely that targeting epigenetic elements alone will suffice in treating all types of cancer.

Proteomics analyzes cellular proteins, the translational products of RNA transcripts, and mediators of cellular functions by quantifying peptide abundance, modifications, and interactions [31]. Mass spectrometry

(MS) has been adapted for high-throughput analyses of thousands of proteins, facilitating the identification of global proteome interactions and quantification of posttranslational modifications, ultimately characterizing protein pathways, networks, and their overall effects on the system [31]. However, proteomic studies have revealed a low correlation between mRNA expression and the corresponding protein expression levels, owing to extensive posttranslational processing, which requires additional analytical methodologies, such as nuclear magnetic resonance (NMR) and X-ray crystallography, to decipher [5]. Proteomic research has described several cancer-specific biomarkers, which can be used in classifying cancer types and predicting drug sensitivity and resistance [32–34].

Metabolomics focuses on quantifying numerous small molecules and products of cellular metabolism, such as amino acids, fatty acids, and carbohydrates. Metabolite levels and relative ratios reflect metabolic activity, which may be dysregulated in diseased states [35]. MS-based techniques are employed to quantify relative and targeted small molecule abundances [5]. Although studies attempted to identify tumor-specific biomarkers in plasma or serum samples in the hope of removing the need for invasive biopsy samples, such plasma/serum metabolites are highly nonspecific and, therefore, of limited significance. Nevertheless, research has revealed altered carbohydrate metabolism in acute myeloid leukemia and dysregulated unsaturated free fatty acid metabolism in colorectal cancer [36,37].

Microbiomics studies microbial cells (bacteria, fungi, protozoa, and viruses) that colonize the epithelial surfaces of humans. These microorganisms and their genetic material collectively compose the microbiome, a highly complex system [38]. Multiple studies have implicated perturbations in gut microbiota in a variety of diseases [39,40]. Microbiomic profiling utilizes amplification and sequencing of specific hypervariable 16S rRNA genes, which allows clustering into operational taxonomic units. Shotgun metagenomics utilizes total DNA sequencing, which provides additional resolution for distinguishing genetically similar species. NGS of 16S ribosomal abundance and metagenomics quantifications allow the determination of the microbiome and facilitate correlation with phenotypic states of interest. However, the direction of causality is unclear, as it remains elusive whether microbiome profiles are a cause or the product of diseased states.

The study of "omics" can be extended to a wide range of processes. Lipidomics builds on the principles of proteomics and metabolomics, focuses on the study of organic solvent-soluble molecules, and aims at establishing the cell's cellular metabolic history and associated functional profile [41]. Similarly, extending from proteomic studies, glycomics examines the glycosylation processes and provides insights into the functional diversity that generates distinct phenotypes from a limited genotype [42]. Foodomics studies the bioactive molecules and mechanisms associated with nutrition and investigates the overall effects on human health [43]. Furthermore, cellomics is defined as the creation of mediums that enable cellular processes to occur and allow cellular relationships to develop, facilitating the study of cellular behaviors, such as cell migration, differentiation, death, etc. [44]. This principle can also be expanded in the study of other microorganisms, as seen in metagenomics. Metagenomics constitutes culture-independent analyses of microorganisms, including diverse organisms that cannot be cultured by conventional means [45]. Notably, "omic" reasoning can also be applied in remarkably more complicated functions such as cognition. For example, connectomics attempts to discern anatomical patterns and functional matrixes that exist within the human brain and provide insight into cognitive processes [46]. A detailed discussion of these "omics" is beyond the scope of this chapter.

"Multiomics" applications in precision medicine

Each type of "omic" data provides insight into which biological pathways and functions may be involved in disease processes. However, examination of only one type of "omic" data is limited to correlation, as mostly reactive associations are captured rather than causative interactions. Different "omic" methodologies assess different parts of the pathophysiological mechanism, with little overlap between them [5]. Therefore, integrating different types of "omics" provides a holistic perspective, establishes causative changes, and expedites the development of predictive and treatment models by discerning the genotype-to-phenotype relationship. Implementation of multiomic data creates informatics and interpretation challenges. Combining various "omic" data sets demands novel analytical approaches and statistical methods with high reliability and actionability to guide clinical care. Below we describe examples of "multiomics" applications in precision medicine.

Multiomic analyses have been conducted in Mendelian diseases and, in 25%—50% of the cases, have been able to identify causative mutations, which have not been previously included in targeted gene panels [47—49]. When the effects of a disease-causing variant are subtle, difficult to detect, or not previously well described, identification of a causal relationship is even more challenging. Integrating "multiomic" information, such as RNA sequencing, enhances the detection of key molecular events that may strengthen the association with a specific variant. For instance, in a study on Fanconi anemia, genomic analysis of DNA sequencing and array comparative genomic hybridization (aCGH) detected causal mutations. Similarly, RNA-sequencing (RNA-seq) identified certain previously unknown pathogenic variants, which resulted in decreased expression of a transcript, complementing the information extrapolated from the genomic analysis [50]. Another study on muscular dystrophy combined genomic whole exome sequencing (WES) and transcriptomic RNA-seq analysis to establish the diagnosis [51]. WES did not reveal any suspicious candidates, but RNA-seq identified splice abnormalities. Interestingly, the candidate variants were intronic and otherwise not predicted to cause a downstream effect; therefore, even GWAS may not have detected them. These examples underlie the need to combine multiple "omic" data types to obtain a complete and representative reflection of disease pathophysiology.

Oncological applications of "multiomic" analyses have significantly impacted cancer profiling, diagnosis, and treatment. Combining traditional genomic information with additional "omic" data facilitates the identification of molecular mechanisms underlying disease-causing genetic variability. For example, WES, copy number variation microarray, and RNA-seq identified a fusion gene (EGFR-SEPT14), which was functionally validated to participate in glioma growth [52]. Recruiting multiple "omic" data and complementing genetic analyses resulted in mechanistic explanations that functionally link the relevant causal genetic variants and enable prioritization of the variants. Furthermore, applying a "multiomic" approach provides insight into dysregulated general biochemical pathways at different levels, exposing different types of information. For instance, research has shown that proteomic profiles do not perfectly correlate with transcriptomic and genomic data, suggesting that noncoding mutations or indirect pathways are implicated [53,54]. As such, studies have revealed several noncoding regions that are involved in oncogenesis and highlight the need to integrate multiple "omic" information, to obtain a comprehensive understanding of the underlying pathways.

In addition, "multiomic" integration can provide insights into disease processes that are shared across individual patients. Recent publications have described molecular pathophysiological processes in Alzheimer's disease (AD) and obesity. Gene expression and epigenomic data revealed upregulated immune cell enhancer signatures. Although associations between immune genes and AD have been well documented, recruiting multiple "omic" methodologies clarified the direction of effect [55]. Likewise, epigenomic, transcriptomic, and chromosomal conformation analyses in obese patients with the FTO allele proposed a mechanistic model for the risk allele, which, when restored, reversed aberrant expression and thermogenesis, proposing a potential therapeutic target [56]. These studies provide information about pathophysiological processes that are shared in a diseased state across individuals.

Furthermore, synthesizing data from multiple layers of biological mechanisms, and characterizing "multiomic" profiles of individuals over time, can provide great insight into the molecular effects that lead to physiological phenotypes. Chen et al. conducted transcriptomic, proteomic, and metabolomic analyses, identifying potential suspicious genetic variants in developing type 2 diabetes mellitus [57]. The candidate variants are known to participate in insulin signaling and were shown by transcriptomic and proteomic data to be downregulated during a viral infection, which correlated with increased blood glucose concentration. The benefit of such analyses lies in their ability to control for genetic and individual background, therefore tracking mechanistic relationships across multiple "omic" layers and associating them with phenotypic outcomes.

At present, only a handful of "multiomic" practices have been proven to outperform traditional clinical tests, and consequently, significant technical and regulatory barriers hinder the incorporation of these techniques into clinical practice. With the rapid technological advancement, clinical integration of "multiomic" approaches will lead to profound insights into diseased states and ultimately improve patient outcomes.

Challenges

Multiomic profiling presents a cost-effective and practical approach to detecting comprehensive large-scale, system-wide change. Ideally, conducting such longitudinal profiling will establish patient-specific trends that will guide clinical practices and ultimately enhance precision medicine. However, achieving this goal requires practical and innovative solutions to current biological, technological, and analytical challenges.

The field of integrative omics poses several analytical challenges that hinder its widespread clinical implementation. Notably, a robust and reproducible analytical framework is fundamental to analyzing multiple discrete data sets, each unique in its characteristics and biases [58]. Therefore, multistage or multidimensional analysis is required, which demands significantly sophisticated methods, such as neural networks, Bayesian models, and dimensionality reduction [59,60]. This complexity is further convoluted by the distinct nature and, consequently, different types of data obtained from each type of omic profiling. For example, genomic information is expressed by discrete and static metrics, whereas transcriptomic information, such as RNA-seq, is represented by continuous variables [58]. This complexity is also propagated by our inability to discern reactive from causative changes, limiting us to making mainly correlative inferences. In theory, identifying the variation driving the phenotypic changes and also establishing the temporal association between variant and trait embodies the first step forward. Even though this correlative nature is less pronounced in genomics, the final proof of causation linking omic data to phenotypic changes remains elusive.

Furthermore, most analytical methods are mainly focused on shedding light on biological and disease processes and are not specifically designed to be employed on the patient level within a clinical context. To bridge this gap, multiple population data sets are required [61]. Inherently, analyzing and storing such data is associated with enormous computational needs. For reference, genomic profiling of an individual generates data in the terabyte range, and a database, including thousands of patients, requires exabytes of storage. Consequently, the associated computational infrastructure and storage costs far exceed the capabilities of classic research laboratories. Fortunately, cloud-computing-based solutions may eliminate the need for in-house infrastructure and promote reproducibility in computational processes [62]. Such approaches, like the Galaxy project, hold promising potential in handling and processing high-throughput data.

Creating reliable, high-quality cohorts that accurately capture areas of variance within the relative population are fundamental. Technological advancement has rendered this collection of data feasible. Interrogating the population in question at multiple omic levels, as well as in various environmental conditions and diverse backgrounds, has been proven effective. For example, studies such as the MuTher study and the Metabolic Syndrome In Men evaluated multiomic data in different populations, which yielded significant insights into the genetic control of various molecular processes and the metabolic pathways associated with various phenotypes, including cardiovascular traits [63,64]. Streamlining high-quality data collection requires a robust framework for incorporating such information in electronic patient records.

Apart from the infrastructure and computational challenges, appropriate end-user training is critical. The scientist conducting the analyses is required to (1) recognize the problem at hand, (2) discern the intricacies associated with the respective data analyses, and (3) understand the advantages and disadvantages associated with the available computational platforms and draw accurate and sound inferences [58].

As is the case for any technology, implementing "multiomic" analyses as part of regular clinical practice necessitates high specificity and sensitivity. Currently, except for WES and WGS in very particular instances, such technologies have failed to demonstrate superiority to current practices. Progressing forward, to guide clinical protocols and facilitate personalized medicine, research establishing accuracy and efficacy, as well as demonstrating noninferiority and cost-effectiveness, is necessary.

However, barriers between various disciplines participating in the study of "multiomics" pose further hindrances. The fields of informatics, mathematics, statistics, biology, and medicine have been developed on divergent backgrounds, and to establish links between them, a high threshold for a mixture of signals and errors has to be tolerated [62]. The new generation of scientists, who will be called to participate in "multiomic" analyses, will be required to work within multidisciplinary teams and develop a concrete understanding of the various elements involved. Integrating and incorporating "omic" data constitutes a critical limiting factor in translating "multiomics" into personalized medicine [65]. Overcoming such obstacles may boost the nationwide implementation of "multiomics" and personalized medicine and potentially overcome the limitations of two-tiered medical systems worldwide [62].

Conclusions

Overall, it is evident that the concept of precision and personalized medicine is multifaceted and requires much more than combining multiple "omic" approaches. As research continues and technology evolves, scientists need to keep generating comprehensive, unbiased, and truly "multiomic" databases, which are necessary to develop the analytical, statistical, computational, and biological tools required to transition to precision medicine. Untangling these extremely complicated mechanisms constitutes a time- and resource-consuming process. We envision that the study of "multiomics" will yield profound insights into human disease, which will eventually be integrated into decision-making and become commonplace in future clinical practice.

References

[1] Donovan C. Biological function. In: Machluf K, editor. Encyclopedia of evolutionary psychological science. Springer; 2019. Accessed 25 October 2022.

[2] Krzyszczyk P, Acevedo A, Davidoff EJ, et al. The growing role of precision and personalized medicine for cancer treatment. Technol (Singap World Sci) 2018; 6(3−4):79−100.

[3] Nass CM, Omenn SJ. GS CotRoO-BTfPPOiCTBoHCSBoHSPIoMM. Evolution of translational omics: lessons learned and the path forward. Washington, DC: National Academies Press; 2012.

[4] Yadav SP. The wholeness in the suffix -omics, -omes, and the word om. J Biomol Tech 2007;18(5):277.

[5] Olivier M, Asmis R, Hawkins GA, Howard TD, Cox LA. The need for multiomics biomarker signatures in precision medicine. Int J Mol Sci 2019;20(19).

[6] Crick F. Central dogma of molecular biology. Nature. 1970;227(5258):561−3.

[7] Uffelmann E, Huang QQ, Munung NS, de Vries J, Okada Y, Martin AR, et al. Genome-wide association studies. Nat Rev Methods Prim 2021;1:59.

[8] Silva FC, Valentin MD, Ferreira Fde O, Carraro DM, Rossi BM. Mismatch repair genes in Lynch syndrome: a review. Sao Paulo Med J 2009;127(1):46−51.

[9] Kurian AW. BRCA1 and BRCA2 mutations across race and ethnicity: distribution and clinical implications. Curr Opin Obstet Gynecol 2010;22(1):72−8.

[10] Alexandrov LB, Nik-Zainal S, Wedge DC, et al. Signatures of mutational processes in human cancer. Nature 2013;500(7463):415−21.

[11] Guha T, Malkin D. Inherited TP53 mutations and the Li-Fraumeni syndrome. Cold Spring Harb Perspect Med 2017;7(4).

[12] Ben-Neriah Y, Daley GQ, Mes-Masson AM, Witte ON, Baltimore D. The chronic myelogenous leukemia-specific P210 protein is the product of the bcr/abl hybrid gene. Science 1986;233(4760):212−14.

[13] Bailey MH, Tokheim C, Porta-Pardo E, et al. Comprehensive characterization of cancer driver genes and mutations. Cell 2018;173(2):371−85 e318.

[14] Bolton KL, Chenevix-Trench G, Goh C, et al. Association between BRCA1 and BRCA2 mutations and survival in women with invasive epithelial ovarian cancer. JAMA 2012;307(4):382−90.

[15] Gao Q, Liang WW, Foltz SM, et al. Driver fusions and their implications in the development and treatment of human cancers. Cell Rep 2018;23(1):227−38 e223.

[16] Davoli T, Uno H, Wooten EC, Elledge SJ. Tumor aneuploidy correlates with markers of immune evasion and with reduced response to immunotherapy. Science 2017;355(6322).

[17] Lee E, Iskow R, Yang L, et al. Landscape of somatic retrotransposition in human cancers. Science 2012;337(6097):967−71.

[18] Morozova O, Hirst M, Marra MA. Applications of new sequencing technologies for transcriptome analysis. Annu Rev Genomics Hum Genet 2009;10:135−51.

[19] Nica AC, Dermitzakis ET. Expression quantitative trait loci: present and future. Philos Trans R Soc Lond B Biol Sci 2013;368(1620):20120362.

[20] Cancer Genome Atlas Research N, Weinstein JN, Collisson EA, et al. The cancer genome atlas pan-cancer analysis project. Nat Genet 2013;45(10):1113–20.

[21] Robinson D, Van Allen EM, Wu YM, et al. Integrative clinical genomics of advanced prostate cancer. Cell 2015;161(5):1215–28.

[22] Botling J, Edlund K, Lohr M, et al. Biomarker discovery in non-small cell lung cancer: integrating gene expression profiling, meta-analysis, and tissue microarray validation. Clin Cancer Res 2013;19(1):194–204.

[23] Duarte CW, Willey CD, Zhi D, et al. Expression signature of IFN/STAT1 signaling genes predicts poor survival outcome in glioblastoma multiforme in a subtype-specific manner. PLoS One 2012;7(1):e29653.

[24] Prat A, Ellis MJ, Perou CM. Practical implications of gene-expression-based assays for breast oncologists. Nat Rev Clin Oncol 2011;9(1):48–57.

[25] Suva ML, Tirosh I. Single-cell RNA sequencing in cancer: lessons learned and emerging challenges. Mol Cell 2019;75(1):7–12.

[26] Song Q, Hawkins GA, Wudel L, et al. Dissecting intratumoral myeloid cell plasticity by single cell RNA-seq. Cancer Med 2019;8(6):3072–85.

[27] Kelsey G, Stegle O, Reik W. Single-cell epigenomics: recording the past and predicting the future. Science 2017;358(6359):69–75.

[28] Campagna MP, Xavier A, Lechner-Scott J, et al. Epigenome-wide association studies: current knowledge, strategies and recommendations. Clin Epigenetics 2021; 13(1):214.

[29] Hegi ME, Diserens AC, Gorlia T, et al. MGMT gene silencing and benefit from temozolomide in glioblastoma. N Engl J Med 2005;352(10):997–1003.

[30] El-Awady RA, Hersi F, Al-Tunaiji H, et al. Epigenetics and miRNA as predictive markers and targets for lung cancer chemotherapy. Cancer Biol Ther 2015; 16(7):1056–70.

[31] Graves PR, Haystead TA. Molecular biologist's guide to proteomics. Microbiol Mol Biol Rev 2002;66(1):39–63 table of contents.

[32] Swiatly A, Horala A, Matysiak J, Hajduk J, Nowak-Markwitz E, Kokot ZJ. Understanding ovarian cancer: iTRAQ-based proteomics for biomarker discovery. Int J Mol Sci 2018;19(8).

[33] Cruz IN, Coley HM, Kramer HB, et al. Proteomics analysis of ovarian cancer cell lines and tissues reveals drug resistance-associated proteins. Cancer Genomics Proteom 2017;14(1):35–51.

[34] Ali M, Khan SA, Wennerberg K, Aittokallio T. Global proteomics profiling improves drug sensitivity prediction: results from a multi-omics, pan-cancer modeling approach. Bioinformatics 2018;34(8):1353–62.

[35] Clish CB. Metabolomics: an emerging but powerful tool for precision medicine. Cold Spring Harb Mol Case Stud 2015;1(1):a000588.

[36] Chaturvedi A, Araujo Cruz MM, Jyotsana N, et al. Mutant IDH1 promotes leukemogenesis in vivo and can be specifically targeted in human AML. Blood 2013; 122(16):2877–87.

[37] Zhang Y, He C, Qiu L, et al. Serum unsaturated free fatty acids: a potential biomarker panel for early-stage detection of colorectal cancer. J Cancer 2016; 7(4):477–83.

[38] Nkera-Gutabara CK, Kerr R, Scholefield J, Hazelhurst S, Naidoo J. Microbiomics: the next pillar of precision medicine and its role in African healthcare. Front Genet 2022;13:869610.

[39] Kostic AD, Xavier RJ, Gevers D. The microbiome in inflammatory bowel disease: current status and the future ahead. Gastroenterology 2014;146(6):1489—99.

[40] Turnbaugh PJ, Ley RE, Mahowald MA, Magrini V, Mardis ER, Gordon JI. An obesity-associated gut microbiome with increased capacity for energy harvest. Nature 2006;444(7122):1027—31.

[41] Han X, Gross RW. Shotgun lipidomics: multi-dimensional MS analysis of cellular lipidomes. Expert Rev Proteom 2005;2(2):253—64.

[42] Raman R, Raguram S, Venkataraman G, Paulson JC, Sasisekharan R. Glycomics: an integrated systems approach to structure-function relationships of glycans. Nat Methods 2005;2(11):817—24.

[43] Braconi D, Bernardini G, Millucci L, Santucci A. Foodomics for human health: current status and perspectives. Expert Rev Proteom 2018;15(2):153—64.

[44] Primiceri E, Chiriaco MS, Rinaldi R, Maruccio G. Cell chips as new tools for cell biology—results, perspectives and opportunities. Lab Chip 2013;13(19):3789—802.

[45] Riesenfeld CS, Schloss PD, Handelsman J. Metagenomics: genomic analysis of microbial communities. Annu Rev Genet 2004;38:525—52.

[46] Sporns O, Tononi G, Kotter R. The human connectome: a structural description of the human brain. PLoS Comput Biol 2005;1(4):e42.

[47] Jacob HJ, Abrams K, Bick DP, et al. Genomics in clinical practice: lessons from the front lines. Sci Transl Med 2013;5(194):194cm195.

[48] Lee H, Deignan JL, Dorrani N, et al. Clinical exome sequencing for genetic identification of rare Mendelian disorders. JAMA 2014;312(18):1880—7.

[49] Yang Y, Muzny DM, Reid JG, et al. Clinical whole-exome sequencing for the diagnosis of mendelian disorders. N Engl J Med 2013;369(16):1502—11.

[50] Chandrasekharappa SC, Lach FP, Kimble DC, et al. Massively parallel sequencing, aCGH, and RNA-Seq technologies provide a comprehensive molecular diagnosis of Fanconi anemia. Blood 2013;121(22):e138—48.

[51] Cummings BB, Marshall JL, Tukiainen T, et al. Improving genetic diagnosis in Mendelian disease with transcriptome sequencing. Sci Transl Med 2017;9(386).

[52] Frattini V, Trifonov V, Chan JM, et al. The integrated landscape of driver genomic alterations in glioblastoma. Nat Genet 2013;45(10):1141—9.

[53] Zhang B, Wang J, Wang X, et al. Proteogenomic characterization of human colon and rectal cancer. Nature 2014;513(7518):382—7.

[54] Mertins P, Mani DR, Ruggles KV, et al. Proteogenomics connects somatic mutations to signalling in breast cancer. Nature 2016;534(7605):55—62.

[55] Gjoneska E, Pfenning AR, Mathys H, et al. Conserved epigenomic signals in mice and humans reveal immune basis of Alzheimer's disease. Nature 2015;518(7539):365—9.

[56] Claussnitzer M, Dankel SN, Kim KH, et al. FTO obesity variant circuitry and adipocyte browning in humans. N Engl J Med 2015;373(10):895—907.

[57] Chen R, Mias GI, Li-Pook-Than J, et al. Personal omics profiling reveals dynamic molecular and medical phenotypes. Cell 2012;148(6):1293—307.

[58] Karczewski KJ, Snyder MP. Integrative omics for health and disease. Nat Rev Genet 2018;19(5):299—310.

[59] Holzinger ER, Dudek SM, Frase AT, Pendergrass SA, Ritchie MD. ATHENA: the analysis tool for heritable and environmental network associations. Bioinformatics 2014;30(5):698—705.

[60] Fridley BL, Lund S, Jenkins GD, Wang L. A Bayesian integrative genomic model for pathway analysis of complex traits. Genet Epidemiol 2012;36(4):352—9.

[61] Ota M, Fujio K. Multi-omics approach to precision medicine for immune-mediated diseases. Inflamm Regen 2021;41(1):23.

[62] Alyass A, Turcotte M, Meyre D. From big data analysis to personalized medicine for all: challenges and opportunities. BMC Med Genomics 2015;8:33.

[63] Laakso M, Kuusisto J, Stancakova A, et al. The metabolic syndrome in men study: a resource for studies of metabolic and cardiovascular diseases. J Lipid Res 2017; 58(3):481–93.

[64] Nica AC, Parts L, Glass D, et al. The architecture of gene regulatory variation across multiple human tissues: the MuTHER study. PLoS Genet 2011;7(2):e1002003.

[65] Van den Bulcke TL, Karen, Van de Peer Y, Marchal K. Inferring transcriptional networks by mining 'omics' data. Curr Bioinforma 2006;1(3).

CHAPTER 10

Global impact and application of Precision Healthcare

Alexios-Fotios A. Mentis[1],* and Longqi Liu[2],*
[1]BGI Genomics, Shenzhen, China
[2]BGI Research, Hangzhou, Hangzhou, China

Introduction: *"In the beginning was the Word"*

"Precision Healthcare"—both as a concept and as an implemented approach to healthcare—has received exponentially increasing attention during the last ten years, notably after the flourishing, ripe fruits of the Human Genome Project and its associated translational research. Historically, Precision Healthcare represents the continuum of *Precision Medicine*, which emerged as a term after Personalized or Individualized Medicine, and it signifies a major change in thinking for modern healthcare practice. Nonetheless, Precision Healthcare appears to be the most appropriate terminology for several reasons: (1) The term Precision Healthcare is broader than Precision Medicine, similar to how the term Healthcare is broader than Medicine, and it thus represents a more *patient-centric* approach rather than a *physician-centric* approach [1]; (2) the term *Individualized Medicine* in lieu of *Personalized Medicine* may stimulate potential philosophical debates, as the focus of humanistic ethics has been the *person* as a whole psychosomatic entity and not the individual as a single unit [2], let alone the gradual replacement of its cognitive functions by algorithmic decisions (described also as *"losing the patient from the picture"* [3]); and (3) reaching a fully personalized diagnostic and therapeutic approach may be, more or less, utopic at least for the forthcoming period, given both the complexity of the pathophysiological underpinnings of every single human being (and potential disease candidate) and the need to inadvertently follow and hopefully align to the guidelines of evidence-based medicine during current clinical practice [4].

Here, we will provide an overview of how Precision Healthcare can be applied on a global scale and leave its impact on 21st century medicine. Initially, we will focus on recent genomics breakthroughs (i.e., advanced high-throughput sequencing) that have the potential to reshape the field of

*Equal contribution.

The New Era of Precision Medicine
DOI: https://doi.org/10.1016/B978-0-443-13963-5.00001-7

© 2024 Elsevier Inc.
All rights reserved.

medical and healthcare knowledge in a way similar to the revolution that the microscope brought to the fields of microbiology and pathology (the *"mother sciences"* of investigating infectious diseases and neoplastic lesions, respectively). We will then provide an overview of the *"N-of-1"* studies that are an indispensable tool of Precision Healthcare across different clinical settings. We will also analyze the challenges faced when applying translational findings into clinical practice and when aiming for equity in the Precision Healthcare era. In addition, we will explore *"neglected"* forms of medicine that have displayed principles of the *"personalized"* element of healthcare and that could be considered complementary to Precision Healthcare. Last, we will discuss how the reproducibility of research findings should be a cornerstone of Precision Healthcare investigations.

"Nth-generation" sequencing approaches: a deep dive into Precision Healthcare

"Molecular medicine" can be perceived as the *"mother"* concept of the 2000s decade that then gave birth to Precision Healthcare. Molecular Medicine has been characterized as an innovative approach to examining life and cure pathology at the molecular and cellular level by providing additional levels of mechanistic understanding compared to the phenotypically driven clinical practice [5]. During the last decade, a new term, that of *"High-Definition Medicine,"* has emerged in the scientific literature, to describe *"the dynamic assessment, management, and understanding of an individual's health measured at (or near) its most basic units,"* using the synergistic combination of several diagnostic and therapeutic tools, from germline DNA sequencing, circulating multi-*omics* (i.e., DNA, RNA, methylation, metabolites, and so on) sequencing, imaging, and behavioral tracking up to cellular therapies, artificial intelligence, and microbiome among others [6].

For Precision Healthcare to reach its fullest potential, it is essential to dive into the most fundamental units of life, i.e., the single cell (or even deeper) for a number of distinct reasons. These include the large heterogeneity, not only between different people or between different tissues of the same body (as these questions can already be addressed by bulk sequencing) but also within the same tissue. Single-cell genomics and, more broadly, single-cell *"omics"* aim to address this major pending gap by studying the genetic material (and its expression) on a single-cell level. Defining new cell types and understanding the intercellular dynamics are key aspects that can be addressed through such high-throughput approaches [7,8].

Nevertheless, and despite the astonishingly promising results of such techniques, the need for another level of *"High-definition"* medicine appeared: How is the expression of genetic information organized in each cell in a space- and time-wise manner? This question has been the chief one leading several research groups and next-generation sequencing technologies to develop and perform experimentation on the so-called *"spatial transcriptomics"* that was voted by the journal *"Nature Methods"* as *"Method of the year in 2020"* [9−11]. Thus, it should come as no surprise if future studies refer to *N*th-generation sequencing approaches, besides the current third generation of long-read sequencing [12], to reflect the continually deepening levels of biological precision reached by such high-throughput approaches.

Intriguingly, single-cell and spatial transcriptomics approaches have already yielded several ripe fruits. To begin with, single-cell approaches have provided the most precise cellular atlas of the primate *Macaca fascicularis* to date, encompassing 45 tissues and one million cells [13]. This breakthrough becomes particularly significant when considering the undeniable extrapolations of such findings to humans, and especially with clinical relevance, such as how some viruses may have tissue specificity [13]. In addition, applications of spatial transcriptomics include the creation of spatiotemporal atlases for all experimental models, which have traditionally been pivotal in advancing life sciences and biomedical research, from *Arabidopsis thaliana* [14] and *Drosophila* species [15] to zebrafish [16] and mouse [17]. Moving forward, future studies could focus on assessing tumor heterogeneity at the intraindividual, and not only the interindividual level, as well as on performing functional genomics studies that—through the multidimensional (*"spatiotemporal"*) component—could allow *"charting"* the developmental effects of specific clinically significant variants into model organisms [18]. However, ensuring that these techniques will reach clinical prime time would require tenacious efforts given the several challenges that must be overcome; in particular, the low histological quality of several clinical specimens leading to low RNA integrity scores [19].

Collectively, such approaches are in complete alignment with the goal set by other nation's official entities, such as the United States National Academies of Science, Engineering, and Medicine, that urged to *"define diseases more precisely"* and, in doing so, to provide a *"new taxonomy of human disease"* [20].

The emerging value of *"N-of-1"* studies

Evidence-based medicine has long supported the hierarchy of evidence through a pyramid-based ranking, with case reports/case series providing the lowest, and clinical trials and systematic reviews/metaanalyses providing the highest level of evidence [21] (even though several suggestions following the initial model have later been published [22–26]). Within this context, recent approaches have moved a step forward to provide a more robust level of evidence: i.e., that of umbrella [27], overviews of reviews [28], or even *"meta-umbrella"* [29] systematic reviews. These latter approaches aim to provide an overarching method of summarizing the hierarchically highest evidence; however, the implementation of such findings into public health policy remains a pending yet pressing gap.

Nonetheless, the history of modern medicine has shown that major breakthroughs have come even from case reports and case series, such as the HIV/pandemic that was initially described on a small group of patients suffering from opportunistic infections [30]. Indeed, case reports and case series may reflect the extreme end of a *"nosological continuum"* that can provide major insight into pathophysiological mechanisms, in turn forming the basis (as already done) for drug development. For example, the development of statins that lower low–density lipoprotein cholesterol levels in the blood in order to reduce atherosclerosis was based on studying cases of familial hypercholesterolemia, where the role of the LDL receptor was discovered [31]. More recently, there has also been the concept of *"molecular case reports,"* in which both functional and computational genomics approaches are used to elucidate the potentially pathogenic role of a specific gene's variant. Of note, such approaches have been applied to a variety of clinical scenarios, ranging from cancer [32] to neurodegenerative disorders [33] or even to infectious and immune disorders [34], and they are perceived as providing *"rare-to-common"* mechanistic insights [35].

Nowadays, founded on Bayesian statistical concepts, novel statistical genetics approaches have shown that searching for rare variants for common disorders might be a very efficient approach for drug discovery, and as so, several studies harnessing this Bayesian approach have been conducted heralding therapeutically promising results [36]. Within the earlier contexts, a novel approach to clinical trials has been developed, the so–called *"N-of-1"* studies. Symbolizing perhaps the epitome of Precision Medicine, these studies are described by expert groups as *"single-patient crossover trials"* in which experimental drugs are used to treat otherwise incurable chronic conditions [37]. Invigoratingly, a second category of these *"N-of-1"* studies refers to individualized treatments for patients suffering from significantly rare or even ultrarare conditions [37]. However, caution has been advised by leading experts on the need to regulate and monitor the methodological quality and scientific robustness of these otherwise very promising clinical trial study design and therapeutic approaches [37].

A pressing moral question though arises: How likely is it that *"N-of-1"* studies can reach populations with limited access to medicines and healthcare? More broadly, if *"social pharmacology,"* which deals with how socially and financially vulnerable groups can reach appropriate treatment standard, might have failed to deliver its mission [38], how likely will that mission be fulfilled by Precision Healthcare? The modern history of healthcare has shown that modern-day clinical practice was part only of elite academic centers in the past. In other words, democratizing access to healthcare and, in this context, adoption of Precision Healthcare may be expected to occur only during the following generations. Besides, it is now commonly observed that the healthcare field appears to lag behind other more fast-forward moving fields like data analytics and software engineering and fintech solutions in speed of adoption and spread of innovation. As a result, approaching Precision Healthcare through the lens of equity should be considered a baseline ethical requirement.

Precision Healthcare: is *"translating"* from *"Bench"* to *"Bedside"* enough to achieve global applications?

At the very core of Precision Healthcare lies the paradigm shift of the so-called *"translational biomedicine"* that took place over the last decades. This shift, mostly pushed forward by the movement of physician-scientists stemming from leading institutions of the era [39,40], also included a shift of pharmaceutical companies toward *"science-driven drug discovery,"* focusing on biological and other compounds that aim to address disease mechanisms, instead of only studying chemically synthesized compounds that were *in vitro, in vivo,* and then clinically evaluated [41]. Nowadays, exciting opportunities for international research networks and consortia on Precision Medicine have been formed, fostering the widespread application of the field on a global scale in lieu of *"oligopolies"* by some high resource-rich countries [42]. While the availability of other-than-genomics approaches (such as RNA sequencing, proteomics, and metabolomics) might have been comparatively limited in the last decade, the genomics-based efforts are still enough to provide clinically meaningful Precision Medicine approaches in a wide spectrum of countries [42]. For instance, countries such as Estonia and Luxemburg have focused on leveraging samples collected from biobanks to address medical needs such as Parkinson's disease or pharmacogenetics-guided treatment. In comparison, others have concentrated their research efforts on genetic diseases with major regional importance, such as gastrointestinal cancers and the Stevens-Johnson syndrome/toxic epidermal necrolysis in Singapore and Thailand, respectively [42].

One of the major culminations of Precision Medicine-intensive efforts to translate bench findings into healthcare consumers, and to individuals who are not strictly *"patients,"* may potentially be contextualized via the so-called *"polygenic risk scores."* These represent a curated, cumulative score for relative risks of a wide spectrum of diseases, based on thousands of single-nucleotide polymorphisms (also known as variants), which are identified by genome-wide association studies (GWAS) and which are, in turn, linked to specific human (patho)physiological traits [43]. Such traits may refer to different cancer types, myocardial infarction, or diabetes, and they are often verified by secondary research studies (notably, GWAS metaanalyses) [44,45]. Despite their increasing popularity, voices of concern have criticized that polygenic risk scores may have clinical validity but still lack clinical utility (for a difference between the two terms, see [46]), especially regarding treatment decisions that require separate validations (a critique captured in the phrase *"prognosis without promise"*) [43]. Similarly, others have stipulated that these scores may favor Caucasian populations (characterized as *"Eurocentric biases"*), ultimately working against equity in Precision Healthcare [47]. Lastly, some reluctance has been expressed, urging for stricter levels of statistical significance in order to establish acceptable standards of predictability at the individual level [48]. This issue becomes more evident amid observations that genetic traits typically, by their nature, have small effects compared to the gross effects of human behavioral health determinants that are resistant to change [49].

Several caveats have been expressed though on the links between translational biomedicine and Precision Medicine. To begin with, the focus on the *"bench-to-bedside"* transition of novel therapeutic compounds and medical devices (among others) whose discovery and/or invention stems from basic science has predominated the concept of *"translational research,"* leaving little space to other *"translational aspects"* [50]. In fact, equally important to the aforementioned is the *"implementation research"* (1) focusing on how patients are better served within the Healthcare system, which is, in turn, judged by the quality of their services, (2) ascertaining that patients have appropriate access to meaningful treatment options, and (3) providing evaluations of medical interventions using real-world data and in settings beyond ones that are hospital based, such as school and workplace, in order to ultimately guide physicians' practice [50]. To this end, applying other sciences beyond basic biomedical research, such as public health, politics, marketing, and behavioral economics, may be crucial [50]. Moreover, the clinical diagnosis and, more broadly, the phenotypic evaluation should remain the cornerstone of Precision Medicine approaches. This has been demonstrated by recent *"molecular case studies,"* which detail how misinterpretation of variants of undetermined significance in probands' relatives has provoked unnecessary and troublesome therapeutic approaches (with a proband being defined as an individual afflicted with a disease who represents the first subject in such a molecular case study). Surprisingly, these could have been avoided if standard clinical diagnostic approaches and genetic molecular autopsies (the latter serving as the equivalent of clinical examination to deceased patients) had been applied sooner [51]. Lastly, we have previously expressed concerns that (1) diseases of resource-poor settings, such as tropical parasitic diseases, may lay behind in Precision Medicine approaches; (2) *"One-size-fits-all"* public health approaches, such as tobacco taxation, may be effective and do not require extensive research funding on *"tailored"* genetic approaches (which could instead be allocated to implementing well-known yet politically neglected policies); and (3) reaching a consensus on what level of *"-omics"* detail is clinically meaningful yet concurrently sustainable in financial terms would require extensive research [3].

Precision Healthcare and equity: the challenge from "Bench to the Agora"

As it is hard to deny that access to healthcare has been historically limited to several population groups, ensuring equity in healthcare and, in turn, in Precision Healthcare is of primordial importance, even though such claims risk being, more or less, wishful thinking in many clinical settings. For example, in the management of diabetes mellitus and related complications—one of the archetypes of internal medicine disorders—the inequities observed are so profound that they leave tight margins for any forthcoming optimism [52]. This is because, on the one hand, the social and financial determinants of health exert such a profound impact on healthcare and access to healthcare systems while, on the other hand, provoking tectonic shifts in these determinants may require several generation's efforts to overcome [53].

In the Precision Healthcare field, inequity becomes more complicated for several reasons, rendering the WHO's call to *"leave no one behind"* and to *"reach the furthest behind first"* even more complex and challenging [54,55]. First of all, the prohibitive cost of high-throughput genomics technologies could make its implementation prohibitive in resource-poor settings, both at the national and subnational level. Secondly, such sophisticated techniques have not fully reached clinical prime time and laboratory flow's automation, thus requiring highly skilled personnel, which may not be available in such settings. Similarly, the interpretation of such Precision Medicine-related findings may require a team of highly specialized healthcare professionals, such as medical geneticists and genetic counselors, whose presence is still limited to only few countries [56].

Another poorly described major area of concern is the unequal representation of different ethnic and racial disparities within genome-wide association and other genomics-inspired studies, which are typically extensively focused on Caucasian and much less on Hispanic or Asiatic populations [57]. Likewise, the degree of *"genomics health literacy"* between populations may be significantly different (as previously shown) [58,59], thus creating a vicious cycle between lack of research populations and lack of genomics-related knowledge. Here lies the bold challenge in reaching *"genomics equity"* [60,61] within the WHO and United Nations goal for Universal Health Coverage, which would involve creating resilient healthcare systems that accommodate access to genomics-inspired diagnostics and therapeutic accesses to as many people as possible and, in doing so, render these services reimbursed by the healthcare systems. Ambitious and utopic as these goals may sound, they can only be considered feasible if appropriate health economics studies—combined with those on development economics—are conducted to serve as a roadmap for implementing Precision Healthcare. Similarly, an additional area of concern relates to how people perceive the *"Nature vs. Nurture"* dilemma across diverse cultural settings. As behavioral mindsets and *"lifestyles"* lie at the core of health determinants [62], the degree to which different cultural, ethnic, and racial groups tend to (1) perceive genomic information in a fatalistic approach (or not) [63] and (2) fear that obtaining genomic knowledge will result in being exposed to genetic discrimination [64] may create obstacles toward a global scenario where humankind benefits from the full potential of Precision Healthcare.

How Precision Healthcare will interact with public health is a crucial yet pending question that needs to be explored. Indeed, it has been noted that while population health stems from improving the health of distinct individuals, public health is population centric at its core, thus considering the in-need-of-improvement whole population as more than the sum of individuals [20]. Borrowing from the tools and concepts of Precision Medicine, the idea of *"precision public health"* was recently proposed [20,65]. According to this concept, both *"-omics"* (i.e., beyond genomics) and advanced data analytics approaches should be leveraged to target population groups that are the most vulnerable and/or those who would benefit most from such public health-oriented interventions [20,65]. In other words, *"precision public health"* has been defined as *"providing the right intervention to the right population at the right time"* [66,67]. Of note, the fundamental tenets behind this principle are (1) the classical economic dogma that resources are always limited, (2) the *"prevention paradox"* as initially expressed by Geoffrey Rose, according to which it might be more efficient to mitigate risks in the comparatively higher number of low-to-average risk patients rather than the comparatively lower number of high-risk patients, as well as (3) robust observations that major differences in progress toward better population health may exist not only between different continents and countries but also within different subnational regions and even neighborhoods [20,65,68—71]. Remarkable examples of *"precision public health approaches"* (as discussed in [49]) include, for example, (1) the national newborn screening program for inborn errors of metabolism and related disorders, such as phenylketonuria, whose lack of detection soon after birth can lead to major clinical manifestations, such as intellectual disability and failure to thrive [48]; (2) universal screening in patients with familial and/or syndromic forms of cancer that are a cost-effective approach, at least in resource-rich settings [72]; and (3) microbial whole-genome sequencing for outbreaks, epidemics, and pandemics surveillance or for antimicrobial resistance monitoring [73].

Collectively, there has been much debate on whether Precision Medicine and precision public health can be *"friends or foes."* Indeed, beyond the conceptual differences between the two regarding the unit of focus (i.e., individuals versus populations), allocating resources to extensive high-throughput approaches for individualized treatment may act in opposition to addressing the fundamental core of disease pathogenesis, which is found in the social, economic, political, financial, and other structural determinants of health [67]. Moreover, others have stipulated that while the advantages of Precision Medicine should be praised (e.g., regarding the success of immunotherapy for certain cancer types or the success of early cancer prevention based on DNA testing [74,75]), the fundamental obstacles in increasing life expectancy are the so-called *"diseases of despair"* (to which alcohol-related pathologies, suicide, and drug overdose are enlisted, and which are linked to low income rates, primary school-only education, and metabolic syndrome [49]), and those related to environmental pollution (e.g., asbestos-related pathologies); both of them are largely neglected by the public health agenda in contrast to spending for novel yet excessively priced biomedicines [76].

All in all, the crux for how future health systems will interact with the life sciences in academia and industry will relate to how efficiently we can move *"from the cell to society"* [67,77] or, if we attempt a similar analogy, *"from the bench to the agora."* To this end, policymakers, not least at the so-called third administrative level (the level of prefectures, just below the state level), should play a reconciliatory role by defining appropriate health strategies that will reach the highest attainable number of individuals in the most efficacious and cost-effective manner [71]. Ultimately, Precision Medicine and precision public health should act collaboratively, both aiming to promote *"the Highest attainable standard of health"* [78]. Nevertheless, a more reconciliatory approach may stem from a recent paradigm shift in the Healthcare sector: the *"Value-Based Healthcare."* This novel concept emphasizes the need to achieve the best possible outcome for the patient where the value stems from comparing patient-related health outcomes to the relevant cost of healthcare delivery. The radical paradigm shift of this concept may lie in rewarding physicians not on a fee-for-services basis but for providing whole-spectrum guidance to patients for reaching a better health status and achieving appropriate management of chronic diseases through an evidence-based medicine approach. As a result, all stakeholders implicated, from patients and providers to payers and suppliers, and even the society-at-large, can benefit when higher healthcare standards are achieved [79,80]. It is, therefore, in this context that Precision Healthcare could play a vital role, improving patient outcomes by offering diagnostic and therapeutic approaches tailored to patients' needs.

Narrative and traditional medicine: *"Neglected"* forms of Precision Healthcare?

Narrative medicine

At the fundamental core of Precision Healthcare lies the key element of *"personalized"* approaches that present as the antipode to the *"one size fits all"* approach of the past and that are in alignment with the emerging concept of *"Systems Medicine"* [81]. Even though these homogenous approaches have proven necessary in the context of drastic public measures to avert epidemics and increase hygiene standards (once the microbial basis of infectious diseases was founded by Pasteur, Koch, and others), it is straightforward to appreciate that the desire to receive *"personalized"* care lies within common human nature. Within the quest though for *"multithousands-"* or *"millions"*-scale projects in our big-data-driven era, the risk of being lost in genomic data and becoming oblivious to the patients' stories should not be forgotten. Besides, as T. S. Elliott (1888–1965), a famous English poet mentioned in the *"The Rock"* published in 1934, *"Where is the wisdom we have lost in knowledge? Where is the knowledge we have lost in information?"*

To address the stories of patients that have remained silent during daily healthcare practices (and are expected to be so during the data-loaded Precision Healthcare era), the movement of *"narrative medicine"* was developed in 2000 by Rita Charon at Columbia University in the USA. According to her very own words, this term is used to *"refer to clinical practice fortified by narrative competence—the capacity to recognize, absorb, metabolize, interpret, and be moved by stories of illness. Simply, it is medicine practiced by someone who knows what to do with stories"* [82]. In particular, Narrative Medicine has been applied to medical genetics, a specialty closely related to Precision Healthcare, in several contexts from congenital skeletal dysplasia to Down's syndrome to Rett syndrome [83–85]. Only the future will tell if related stories will be expressed, written, and heard by patients and/or doctors who have benefited from the solutions and the rewards, respectively, of Precision Healthcare. Until then, we should tenaciously contend that only by pairing biological metrics (e.g., genome) with patient's stories can we reach the full potential of healthcare as the so-called *"art and science of healing"* [86].

Nonetheless, moving from purely evidence-based approaches to giving praise to *"narrative Precision Medicine"* is expected to require an intermediate step, i.e., that of the so-called *"human-important outcomes"* that will be driven by taking into account the interest of patients regarding their outcome (and not only epidemiological parameters, such as mortality and morbidity) [87]. As some of the founders of evidence-based medicine affirm, the decisions are made by people (most likely, treating physicians), not by the evidence *per se* [88]. As a logical corollary, the decisions for clinical intervention should be addressed to people (hospitalized or ambulatory patient) and not to just restoring laboratory parameters to normal.

Collectively, ensuring the human-centric aspect of healthcare persists is an opportunity that should not be missed in the dawn of the Precision Healthcare era.

Traditional Chinese medicine: an opportunity for Precision Medicine toward holistic approaches?

In the data-intensive medicine era, leaving scarce (if any at all) space for traditional medicine may seem, from first glance, a challenging option. Indeed, the major clinical benefits of targeted therapies, especially in the field of oncology, may even create a radical breakthrough, by converting oncology from a tissue-specific (e.g., breast and colorectal) to a tissue-agnostic, targeted molecule-driven medical specialty [89]. However, recent studies have been increasingly showing that traditional medicinal products (e.g., the so-called "*Traditional Chinese Medicine*" [90]) could have adjunctive roles when combined with mainstream pharmaceutical compounds in two ways, by increasing the therapeutic efficacy of a drug and reducing its side effects [91,92]. As the mechanism of this action is still unknown (but suspected to be epigenetic [93]), future research could address the biological underpinnings of such pharmacological additive (or synergistic) actions. After all, combining mainstream pharmaceutical approaches with proven traditional therapeutic regimens can increase regional environmental sustainability and support local agriculture, beyond the principal health effects [94].

More broadly, opinions that *Traditional Chinese Medicine* is conceptually complimentary to Precision Healthcare have been expressed [93]. In particular, the "*holistic*" approach, as well as the "*pattern differentiation*" of Traditional Chinese Medicine, has been analogized to the "*systems medicine*" and "*disease group subclassification*" approaches of Precision Medicine [93]. The authors move forward by claiming that the "*DNA-centric*" approach of modern pathophysiology that is, in turn, founded on the central dogma of biology (i.e., DNA → RNA → Protein) may fail to provide a complete overview of human diseases, thus calling for additional approaches [93]. Indeed, this notion is in line with current intensive efforts to delineate several other "*omics*" of human disease, such as epigenomics [95], metabolomics [96], proteomics [97], lipidomics [98], glycomics [99], and "*exposomics*" [100] to study the epigenome, and the metabolites, proteins, lipids, glycans, and the collective exposome, respectively.

Toward implementing Precision Healthcare: the challenge of reproducibility

The exponentially increasing body of basic, translational, and notably clinical studies has led some to question the quality of evidence surrounding these studies for several reasons, such as different forms of systematic bias and other major methodological errors (e.g., positive result-related reporting bias, data mining to reach statistical significance levels, and small sample sizes), as well as lack of publishing ethics (e.g., not reporting financial or other conflicts of interest) [101−103]. These facts have led several researchers' and editor's initiatives to call for manifestos in favor of *"reproducible science"* and of *"evidence-based medicine for better healthcare"* and to draft *"rules for good research practices"* [104]. These manifestos state that not least through improving transparency, increasing patients' and other stakeholders' participatory role, through harnessing the systematic (and not selective) use of available evidence, and through promoting the values of evidence-based medicine can clinically beneficial and scientifically robust medical knowledge be concurrently produced [4,105]. Interestingly, a *"paradox"* has been described regarding the processes mentioned earlier, as the call for publicly available data to promote methodological accuracy and enhance reproducibility of findings can lead to *"secondary"* findings by other scientists, which will, in turn, provide validation for the primary scientists of the study [106].

Precision Healthcare field cannot be devoid of the same criticisms made of traditional evidence-based medicine, but it also faces additional challenges. Notably, the explosion of new genetic biomarkers may lead to continuous readjustments of genetic *in vitro* testing as well as bioinformatics tools [107]. Thus, the aforementioned would require (1) constantly updated replication of the findings by ensuring consistency of findings between different data sources, as well as (2) the so-called *"targeted repeatability checks"* by other scientists for those discoveries (at the one-digit percentage level) that might reach clinical prime time. The aforementioned holds particular importance for genomics and other -omics approaches, which, in contrast to the analysis of single nucleotide variations, still have a long road ahead to reach >99.9% analytic validity [107].

Similarly, as clinical trials remain the cornerstone of evidence-based medicine, only such study designs can, by their nature, answer the following question (as previously posed): Is the percentage of patients benefiting from Precision Medicine approaches and, as such, characterized as *"superresponders"* truly benefiting from these approaches, or instead are such percentages part of the normal distribution outliers who are expected to exceed the average patients' survival period? [108]. Such questions highlight the key parameter of adequate population sample in order to solidify the *"precision"* component of Precision Medicine [78]. As small population sizes can significantly affect the power of any clinical or basic research study [109], searching for adequate number of patients with X or Y rare disease is most feasible in settings with high overall population rates such as China or India [110].

Additional approaches to refine the methodological and statistical robustness of Precision Medicine have been previously described [111]; for instance, these include (1) *"Toning-down"* the categorical distinctions between *"responders"* and *"nonresponders"* as such distinctions may be partially casual and not purely causal; (2) quitting cut-off points and other dichotomies that may not resonate with real settings (e.g., using the cut-off of exactly 365 days as a one-year response rate); (3) prefer to conduct cross-sectional *"n-of-1"* studies to either achieve multiple measurements of a single treatment or assess multiple treatments over a specific period on the same person; (4) appreciate that other parameters, such as regression to mean or analytical sensitivity issues, may affect clinical trials' results; and (5) acknowledge the delicate interpretation of statistical findings, especially in the context of "individualization," given that an *"X"* risk reduction might refer to either *"Y%"* of participants at 100% of the time or to *"Y times"* for 100% of participants [111].

As shown earlier, reaching clinical prime time through well-designed clinical trials that have no major systematic flaws remains (and it shall remain as such) the *"touchstone"* in the quest of patients' health; thus, ensuring the quality of these trials in the Precision Healthcare era is pivotal. Notwithstanding these goals, the biological complexity of several diseases, notably cancer and neurodegeneration, cannot be dismissed. Indeed, sophisticated *"-omics"* have elucidated both the genetic determinism, as well as the epigenetic and signaling stochasticity of such disease mechanisms. In turn, the latter can be better described in evolutionary terms, including random events and somatic mutations' adaptation that make each individual's cancer genetically unique [112−114]. Therefore, biological unpredictability can play a crucial role in Precision Medicine clinical trials, beyond that of methodological accuracy.

Conclusion

Precision Healthcare presents as much promise to be discovered as hindrances to be overcome. A suitable approach could lie in the *"Omics for all"* approaches [115], which aims to democratize Precision Healthcare for as much of the population as possible. This vision may be the least obscure pathway to forge policies based on the *"leave no one behind"* concept in the 21st century healthcare agenda.

Acknowledgments

AFAM would like to thank Ms. Anna Gkika for her moral support throughout this study.

Conflicts of interest

AFAM and LL are employees of BGI Group. They report no financial or other conflict(s) of interest that could have exercised an influence on this work.

References

[1] Miles A, Mezzich J. The care of the patient and the soul of the clinic: person-centered medicine as an emergent model of modern clinical practice. Int J Pers Centered Med 2011;1:207−22.

[2] Giannaras C. The freedom of morality; 1984.

[3] Mentis A-FA, Pantelidi K, Dardiotis E, Hadjigeorgiou GM, Petinaki E. Precision medicine and global health: the good, the bad, and the ugly. Front Med 2018;5:67.

[4] Heneghan C, et al. Evidence based medicine manifesto for better healthcare. BMJ 2017;357:j2973. Available from: https://doi.org/10.1136/bmj.j2973.

[5] Papavassiliou AG. Molecular medicine: prometheus unbound. Resolving the "enigma" of medicine by re-defining health and disease. Bioessays 2010;32:453. Available from: https://doi.org/10.1002/bies.201090017.

[6] Torkamani A, Andersen KG, Steinhubl SR, Topol EJ. High-definition medicine. Cell 2017;170:828−43. Available from: https://doi.org/10.1016/j.cell.2017.08.007.

[7] Kalisky T, Quake SR. Single-cell genomics. Nat Methods 2011;8:311−14.

[8] Paolillo C, Londin E, Fortina P. Single-cell genomics. Clin Chem 2019;65:972−85.

[9] Williams CG, Lee HJ, Asatsuma T, Vento-Tormo R, Haque A. An introduction to spatial transcriptomics for biomedical research. Genome Med 2022;14:68. Available from: https://doi.org/10.1186/s13073-022-01075-1.

[10] Moffitt JR, Lundberg E, Heyn H. The emerging landscape of spatial profiling technologies. Nat Rev Genet 2022;. Available from: https://doi.org/10.1038/s41576-022-00515-3.

[11] Marx V. Method of the year: spatially resolved transcriptomics. Nat Methods 2021;18:9−14.

[12] Logsdon GA, Vollger MR, Eichler EE. Long-read human genome sequencing and its applications. Nat Rev Genet 2020;21:597−614.

[13] Han L, et al. Cell transcriptomic atlas of the non-human primate *Macaca fascicularis*. Nature 2022;604:723−31. Available from: https://doi.org/10.1038/s41586-022-04587-3.

[14] Xia K, et al. The single-cell stereo-seq reveals region-specific cell subtypes and transcriptome profiling in Arabidopsis leaves. Dev Cell 2022;57:1299−310 e1294.

[15] Wang M, et al. High-resolution 3D spatiotemporal transcriptomic maps of developing Drosophila embryos and larvae. Dev Cell 2022;57:1271–83. Available from: https://doi.org/10.1016/j.devcel.2022.04.006 e1274.

[16] Liu C, et al. Spatiotemporal mapping of gene expression landscapes and developmental trajectories during zebrafish embryogenesis. Dev Cell 2022;57:1284–98 e1285.

[17] Chen A, et al. Spatiotemporal transcriptomic atlas of mouse organogenesis using DNA nanoball-patterned arrays. Cell 2022;185:1777–92. Available from: https://doi.org/10.1016/j.cell.2022.04.003 e1721.

[18] Zou X, et al. From monkey single-cell atlases into a broader biomedical perspective. Life Med 2022;.

[19] Liu X, et al. Clinical challenges of tissue preparation for spatial transcriptome. Clin Transl Med 2022;12:e669.

[20] Chowkwanyun M, Bayer R, Galea S. Precision public health-between novelty and hype. N Engl J Med 2018;379:1398–400.

[21] Guyatt GH, et al. Users' guides to the medical literature: IX. A method for grading health care recommendations. Jama 1995;274:1800–4.

[22] Berliner L., Hanson R., Saunders B. Child physical and sexual abuse: guidelines for treatment; 2004.

[23] Khan KS, ter Riet G, Popay J, Nixon J, Kleijnen J. Stage II conducting the review: phase 5 study quality assessment. Undert Syst Rev Res Effect 2001;2:1–20.

[24] Hayward RS, Wilson MC, Tunis SR, Bass EB, Guyatt G. Users' guides to the medical literature: VIII. How to use clinical practice guidelines A. Are the recommendations valid? JAMA 1995;274:570–4.

[25] Hadorn DC, Baker D, Hodges JS, Hicks N. Rating the quality of evidence for clinical practice guidelines. J Clin Epidemiol 1996;49:749–54.

[26] Atkins D., et al. Grading quality of evidence and strength of recommendations; 2004.

[27] Aromataris E, et al. Summarizing systematic reviews: methodological development, conduct and reporting of an umbrella review approach. Int J Evid Based Healthc 2015;13:132–40. Available from: https://doi.org/10.1097/xeb.0000000000000055.

[28] Lu C, et al. Saffron (Crocus sativus L.) and health outcomes: a meta-research review of meta-analyses and an evidence mapping study. Phytomedicine 2021;91:153699. Available from: https://doi.org/10.1016/j.phymed.2021.153699.

[29] Mentis AA, Dardiotis E, Efthymiou V, Chrousos GP. Non-genetic risk and protective factors and biomarkers for neurological disorders: a meta-umbrella systematic review of umbrella reviews. BMC Med 2021;19:6. Available from: https://doi.org/10.1186/s12916-020-01873-7.

[30] Barré-Sinoussi F, Ross AL, Delfraissy J-F. Past, present and future: 30 years of HIV research. Nat Rev Microbiol 2013;11:877–83.

[31] Brown MS, Radhakrishnan A, Goldstein JL. Retrospective on cholesterol homeostasis: the central role of scap. Annu Rev Biochem 2018;87:783.

[32] Currall BB, et al. Loss of LDAH associated with prostate cancer and hearing loss. Hum Mol Genet 2018;27:4194–203.

[33] Mentis A-FA, et al. A novel variant in DYNC1H1 could contribute to human amyotrophic lateral sclerosis-frontotemporal dementia spectrum. Mol Case Stud 2022;8:a006096.

[34] Bader-Meunier B, et al. Effectiveness and safety of ruxolitinib for the treatment of refractory systemic idiopathic juvenile arthritis like associated with interstitial lung disease: a case report. Ann Rheumatic Dis 2022;81 e20–e20.

[35] Casanova JL, Abel L. From rare disorders of immunity to common determinants of infection: following the mechanistic thread. Cell 2022;185:3086–103. Available from: https://doi.org/10.1016/j.cell.2022.07.004.

[36] Wang Q, et al. Rare variant contribution to human disease in 281,104 UK Biobank exomes. Nature 2021;597:527−32.

[37] Selker HP, et al. A useful and sustainable role for N-of-1 trials in the healthcare ecosystem. Clin Pharmacol Ther 2022;112:224−32.

[38] Papadopulos JS, Mentis A-FA, Liapi C. Social pharmacology as an underappreciated field in medical education: a single medical school's experience. Front Pharmacol 2021;12.

[39] Harding CV, Akabas MH, Andersen OS. History and outcomes of fifty years of physician-scientist training in medical scientist training programs. Acad Med 2017;92:1390.

[40] Daye D, Patel CB, Ahn J, Nguyen FT. Challenges and opportunities for reinvigorating the physician-scientist pipeline. J Clin Invest 2015;125:883−7.

[41] Cockburn IM, Henderson R, Stern S. The diffusion of science-driven drug discovery: organizational change in pharmaceutical research. Cambridge, MA: National Bureau of Economic Research; 1999.

[42] Manolio TA, et al. Global implementation of genomic medicine: we are not alone. Sci Transl Med 2015;7. Available from: https://doi.org/10.1126/scitranslmed.aab0194 290ps213.

[43] Hunter DJ, Drazen JM. Has the genome granted our wish yet? N Engl J Med 2019;380:2391−3. Available from: https://doi.org/10.1056/NEJMp1904511.

[44] Evangelou E, Ioannidis J. Meta-analysis methods for genome-wide association studies and beyond. Nat Rev Genet 2013;14:379−89.

[45] van de Bunt M, et al. Transcript expression data from human islets links regulatory signals from genome-wide association studies for type 2 diabetes and glycemic traits to their downstream effectors. PLoS Genet 2015;11:e1005694.

[46] Burke W. Genetic tests: clinical validity and clinical utility. Curr Protoc Hum Genet 2014;81(9.15.):11−18. Available from: https://doi.org/10.1002/0471142905.hg0915s81.

[47] Martin AR, et al. Clinical use of current polygenic risk scores may exacerbate health disparities. Nat Genet 2019;51:584−91. Available from: https://doi.org/10.1038/s41588-019-0379-x.

[48] El-Hattab AW, Almannai M, Sutton VR. Newborn screening: history, current status, and future directions. Pediatric Clin 2018;65:389−405.

[49] Khoury MJ, Galea S. Will precision medicine improve population health? JAMA 2016;316:1357−8. Available from: https://doi.org/10.1001/jama.2016.12260.

[50] Woolf SH. The meaning of translational research and why it matters. JAMA 2008;299:211−13.

[51] Ackerman JP, et al. The promise and peril of precision medicine: phenotyping still matters most. Mayo Clin Proc 2016;. Available from: https://doi.org/10.1016/j.mayocp.2016.08.008.

[52] Mentis A-FA, Chrousos GP BMC. Endocrine Disorders' collection of articles on "Reducing inequalities in the Management of Endocrine Disorders". BMC Endocr Disord 2022;22(1):96.

[53] Beller E, Hout M. Intergenerational social mobility: the United States in comparative perspective. Future Child 2006;19−36.

[54] Bukhman G, et al. NCDI poverty commission: bridging a gap in universal health coverage for the poorest billion. Lancet 2020;396:991−1044.

[55] Weber H. Politics of 'leaving no one behind': contesting the 2030 Sustainable Development Goals agenda. Globalizations 2017;14:399−414.

[56] Sun L., Liang B., Zhu L., Shen Y., He L. The rise of the genetic counseling profession in China. In: *American Journal of Medical Genetics Part C: Seminars in Medical Genetics*, (Vol. 181, No. 2,. Hoboken, USA: John Wiley & Sons, Inc. p. 170−6.

[57] Mersha TB, Abebe T. Self-reported race/ethnicity in the age of genomic research: its potential impact on understanding health disparities. Hum genomics 2015;9:1−15.

[58] Gupta A, et al. Genomic health literacy interventions in pediatrics: scoping review. J Med Internet Res 2021;23:e26684.

[59] Ogunrin O, Taiwo F, Frith L. Genomic literacy and awareness of ethical guidance for genomic research in Sub-Saharan Africa: how prepared are biomedical researchers? J Empir Res Hum Res Ethics 2019;14:78−87.

[60] Singer PA, Daar AS. Harnessing genomics and biotechnology to improve global health equity. Science 2001;294:87−9.

[61] Khoury MJ, et al. Health equity in the implementation of genomics and precision medicine: a public health imperative. Genet Med 2022;.

[62] Denton M, Walters V. Gender differences in structural and behavioral determinants of health: an analysis of the social production of health. Soc Sci Med 1999;48:1221−35.

[63] Heine SJ, Dar-Nimrod I, Cheung BY, Proulx T. Advances in experimental social psychology, 55. Elsevier; 2017. p. 137−92.

[64] Otlowski M, Taylor S, Bombard Y. Genetic discrimination: international perspectives. Annu Rev Genomics Hum Genet 2012;13:433−54.

[65] Chretien JP, Gaydos JC, Malone JL, Blazes DL. Global network could avert pandemics. Nature 2006;440:25−6. Available from: https://doi.org/10.1038/440025a.

[66] Khoury MJ, Iademarco MF, Riley WT. Precision public health for the era of precision medicine. Am J Preventive Med 2016;50:398−401.

[67] Evangelatos N., Satyamoorthy K., Brand A. 2018;63:433−4. Springer.

[68] Subramanian SV, Lochner KA, Kawachi I. Neighborhood differences in social capital: a compositional artifact or a contextual construct? Health Place 2003;9:33−44.

[69] Tran KB, et al. The global burden of cancer attributable to risk factors, 2010−19: a systematic analysis for the Global Burden of Disease Study 2019. Lancet 2022;400:563−91.

[70] Rose GA, Khaw K-T, Marmot M. Rose's strategy of preventive medicine: the complete original text. Oxford University Press; 2008.

[71] Davey G, Deribe K. Precision public health: mapping child mortality in Africa. Lancet 2017;390:2126−8.

[72] Salikhanov I, et al. Swiss cost-effectiveness analysis of universal screening for Lynch syndrome of patients with colorectal cancer followed by cascade genetic testing of relatives. J Med Genet 2022;59:924−30. Available from: https://doi.org/10.1136/jmedgenet-2021-108062.

[73] Hawken SE, Snitkin ES. Genomic epidemiology of multidrug-resistant Gram-negative organisms. Ann N Y Acad Sci 2019;1435:39−56. Available from: https://doi.org/10.1111/nyas.13672.

[74] Imperiale TF, et al. Multitarget stool DNA testing for colorectal-cancer screening. N Engl J Med 2014;370:1287−97.

[75] Cohen JD, et al. Detection and localization of surgically resectable cancers with a multi-analyte blood test. Science 2018;359:926−30.

[76] Prasad V. Our best weapons against cancer are not magic bullets. Nature 2020;577:451−2.

[77] Brand A. Public health genomics--public health goes personalized? Eur J Public Health 2011;21(1):2−3.

[78] Hunter DJ. The complementarity of public health and medicine − achieving the highest attainable standard of health. N Engl J Med 2021;385:481−4. Available from: https://doi.org/10.1056/NEJMp2102550.

[79] Catalyst N. What is value-based healthcare? NEJM Catalyst 2017;3.

[80] Porter ME. A strategy for health care reform—toward a value-based system. N Engl J Med 2009;361:109−12. Available from: https://doi.org/10.1056/NEJMp0904131.

[81] Zanin M, et al. An early stage researcher's primer on systems medicine terminology. Netw Syst Med 2021;4:2−50.

[82] Charon R. What to do with stories: the sciences of narrative medicine. Can Family Phys 2007;53:1265−7.

[83] Nowaczyk MJ, Carey JC. Narrative medicine: a call to pens. Am J Med Genet A 2013;161a:2117−18. Available from: https://doi.org/10.1002/ajmg.a.36114.

[84] Nowaczyk MJ. Narrative medicine in clinical genetics practice. Am J Med Genet A 2012;158a:1941−7. Available from: https://doi.org/10.1002/ajmg.a.35482.

[85] Knott M, Leonard H, Downs J, Genetic D. The diagnostic odyssey in Rett syndrome: the experience of an Australian family. Am J Med Genet A 2012;158a:10−12. Available from: https://doi.org/10.1002/ajmg.a.34372.

[86] Savel RH, Munro CL. From Asclepius to Hippocrates: the art and science of healing. Am J Crit Care 2014;23(6):437−9.

[87] Jaeschke R, Douketis J, Nowaczyk M, Guyatt G. Human-important outcomes and evidence-based medicine during the coronavirus disease 2019 pandemic. Pol Arch Intern Med 2020;130:714−15.

[88] Haynes RB, Devereaux PJ, Guyatt GH. Physicians' and patients' choices in evidence based practice. BMJ 2002;324(7350):1350.

[89] Mentis A-FA, Grivas PD, Dardiotis E, Romas NA, Papavassiliou AG. Circulating tumor cells as Trojan Horse for understanding, preventing, and treating cancer: a critical appraisal. Cell Mol Life Sci 2020;77:3671−90.

[90] Tang J-L, Liu B-Y, Ma K-W. Traditional Chinese medicine. Lancet 2008;372:1938−40.

[91] Normile D. The new face of traditional Chinese medicine. Science 2003;299:188−90.

[92] Lu C, et al. Chinese medicine as an adjunctive treatment for gastric cancer: methodological investigation of meta-analyses and evidence map. Front Pharmacol 2021;12:797753. Available from: https://doi.org/10.3389/fphar.2021.797753.

[93] Wang WJ, Zhang T. Integration of traditional Chinese medicine and Western medicine in the era of precision medicine. J Integr Med 2017;15:1−7. Available from: https://doi.org/10.1016/s2095-4964(17)60314-5.

[94] Chen S-L, et al. Conservation and sustainable use of medicinal plants: problems, progress, and prospects. Chin Med 2016;11:1−10.

[95] Feinberg AP, et al. Personalized epigenomic signatures that are stable over time and covary with body mass index. Sci Transl Med 2010;2 49ra6-49ra67.

[96] Papadimitropoulos M-EP, Vasilopoulou CG, Maga-Nteve C, Klapa MI. Metabolic profiling. Springer; 2018. p. 133−47.

[97] Cifani P, Kentsis A. Towards comprehensive and quantitative proteomics for diagnosis and therapy of human disease. Proteomics 2017;17:1600079.

[98] Burla B, et al. Ms-based lipidomics of human blood plasma: a community-initiated position paper to develop accepted guidelines1. J Lipid Res 2018;59:2001−17.

[99] Peng W, et al. Clinical application of quantitative glycomics. Expert Rev Proteom 2018;15:1007−31.

[100] Holland N. Future of environmental research in the age of epigenomics and exposomics. Rev Environ Health 2017;32:45−54.

[101] Chalmers I, Glasziou P. Avoidable waste in the production and reporting of research evidence. Lancet 2009;374:86−9.

[102] Ioannidis JP. Why most published research findings are false. PLoS Med 2005;2: e124. Available from: https://doi.org/10.1371/journal.pmed.0020124.

[103] Ioannidis JP. Why most clinical research is not useful. PLoS Med 2016;13: e1002049. Available from: https://doi.org/10.1371/journal.pmed.1002049.

[104] Schwab S., et al. 2022;18:e1010139. Public Library of Science San Francisco, CA.

[105] Munafò MR, et al. A manifesto for reproducible science. Nat Hum Behav 2017;1:0021. Available from: https://doi.org/10.1038/s41562-016-0021.

[106] Hunter DJ, Longo DL. The precision of evidence needed to practice "Precision Medicine". N Engl J Med 2019;380:2472–4. Available from: https://doi.org/10.1056/NEJMe1906088.

[107] Ioannidis JP, Khoury MJ. Improving validation practices in "omics" research. Science 2011;334:1230–2.

[108] Prasad V. Perspective: the precision-oncology illusion. Nature 2016;537:S63. Available from: https://doi.org/10.1038/537S63a.

[109] Button KS, et al. Power failure: why small sample size undermines the reliability of neuroscience. Nat Rev Neurosci 2013;14:365–76. Available from: https://doi.org/10.1038/nrn3475.

[110] Zhao W. China's campaign against rare diseases. Natl Sci Rev 2022;9. Available from: https://doi.org/10.1093/nsr/nwac015 nwac015.

[111] Senn S. Statistical pitfalls of personalized medicine. Nature 2018;563:619–21. Available from: https://doi.org/10.1038/d41586-018-07535-2.

[112] Maley CC, et al. Classifying the evolutionary and ecological features of neoplasms. Nat Rev Cancer 2017;17:605–19. Available from: https://doi.org/10.1038/nrc.2017.69.

[113] Greaves M, Maley CC. Clonal evolution in cancer. Nature 2012;481:306–13. Available from: https://doi.org/10.1038/nature10762.

[114] Lipinski KA, et al. Cancer evolution and the limits of predictability in precision cancer medicine. Trends Cancer 2016;2:49–63. Available from: https://doi.org/10.1016/j.trecan.2015.11.003.

[115] Media N.R.C. Shaping the future of omics for all. 2022. Nature Index, Big 5 Science Nations.

CHAPTER 11

The financial burden of precision medicine

Sufyan Ibrahim[1,2], Karim Rizwan Nathani[1,2] and Mohamad Bydon[1,2]
[1]Department of Neurologic Surgery, Mayo Clinic, Rochester, MN, United States
[2]Mayo Clinic Neuro-Informatics Laboratory, Mayo Clinic, Rochester, MN, United States

Introduction

Precision medicine (PM), also known as personalized medicine, is an emerging approach that aims to provide tailored medical treatments to individual patients based on their genotypic and phenotypic characterizations, environmental, and lifestyle factors [1]. PM essentially describes the best therapeutic strategy for the right patient at the right time, to identify a person's propensity for a disease or to provide timely and targeted prevention. Over the last several decades, a gradual shift toward patient-centered healthcare opened the door for individualized approaches to diagnostics and treatment. PM provides "timely and cost-effective medical solutions to stratified patient subpopulations with predictable outcome margins" [2].

PM includes, but is not limited to, directing healthcare interventions to individuals with a precise and identifiable collection of traits. Most implementations as of now have relied on a single test as a companion diagnostic to target a specific drug to a known subset of patients. PM applications may also include algorithm-based prescription, risk stratification within population screening programs, and the use of genomic-based diagnostics for uncommon hereditary illnesses.

Science has made significant strides in understanding the molecular basis of illness, rapidly developing novel and effective, rationally designed therapies. Since the mapping of the human genome in 2003, the ease, availability, and subsequently the need associated with genetic testing have increased dramatically. Today, there are over 78,000 diagnostic tests available for over 18,500 genes (NCBI. GTR: genetic testing registry, 2023) [3]. In an ironic twist, the success of science is creating a crisis in the affordability of equitable healthcare.

The New Era of Precision Medicine
DOI: https://doi.org/10.1016/B978-0-443-13963-5.00007-8
© 2024 Elsevier Inc.
All rights reserved.
229

The concept of individual-centered medicine is not only transforming the way healthcare is delivered but also is shifting the way how research and development are being undertaken. Currently, gene-based research and development accounts for 20% of pharmaceutical R&D, and the number of customized medications accessible has climbed by 62% since 2012. Furthermore, more than 70% of cancer medications in development now are PMs. Despite its advantages, the technique has been associated with significant financial issues for payers, patients, and clinicians [4].

Many factors influence the cost-effectiveness of targeted therapies, including the prevalence of a certain gene or variant in a population, the accuracy of a test, and the costs of testing and individualized therapy. As a result, patient outcomes may improve, but the cost-effectiveness of PM continues to be unknown [5]. Experts have recently proposed a value-based approach to PM, which entails quantifying the value of PM treatments and establishing their cost-effectiveness in order to guide policy decisions on reimbursement and R&D funding, particularly in solidarity-based health systems.

PM approach holds great promise for improving patient outcomes and reducing healthcare costs, often associated with inappropriate and expensive pharmacological treatments or hospitalizations for severe adverse drug reactions, thereby eventually enabling the most efficient allocation and use of healthcare resources [6]. But PM also poses significant financial challenges for patients, healthcare providers, and payers. In this chapter, we explore the financial burden of PM, including the cost of genetic testing, drug development, and patient access to personalized treatments.

The rising cost of precision medicine

High cost of genomic sequencing and biomarker testing

One of the main drivers of the financial burden of PM is the high cost of genomic sequencing and biomarker testing. There is a pressing need to make genomic medicine, particularly diagnosis for the undiagnosed, accessible to all people, regardless of social class, income, geographic location, or other demographic characteristics. The most potent tool in genomic medicine is clinical whole-genome sequencing (cWGS). PM heavily relies on cWGS for identifying genetic variations that are associated with specific diseases or drug responses [7].

While the cost of genome sequencing has decreased dramatically in recent years, these technologies are still relatively new and not yet widely available, which means that they are not covered by most health insurance plans. According to a 2019 report by the National Institutes of Health (NIH), the cost of WGS, for example, can range from $1000 to $10,000, depending on the type of sequencing, the laboratory, and other factors such as the technology utilized and the volume of samples processed [8]. This cost can be prohibitive for patients who do not have insurance coverage or who have high out-of-pocket costs and could also be out of reach for many healthcare providers.

Similarly, the cost of biomarker testing can range from a few hundred to several thousand dollars, depending on the type of test and the laboratory. Despite the growing economic assessment of biomarker-guided medicines, it is unclear if the current approaches are enough or whether using alternate approaches would provide different cost-effectiveness outcomes. While it is imperative that new biomarker research could cost more at the moment, the economic implications for these tests could potentially be positive in the long run. For example, in the pediatric diabetic population of the United States, a combined strategy of biomarker screening and genetic testing for maturity-onset diabetes of young (MODY) was found to be less expensive than standard care as it increased average quality of life ($+0.0052$ QALY) and decreased costs ($-$191$) per simulated patient relative to the control arm, and the inclusion of cascade genetic testing highlighted the benefits of these PM-based techniques [9]. Similarly, another study, looking into the cost-effectiveness of precision diagnostic testing (PDT) for PM approaches against non-small-cell lung cancer, revealed that when PDT-guided therapy was compared with a therapy-for-all patients approach, all scenarios (100%) proved cost-effective [2].

Widespread adoption of similar strategies might enhance the quality of life for people with certain diseases while saving the healthcare system expenses, demonstrating the potential advantages of personalized medicine for population health. This also suggests that more thorough health economic analysis might reveal other strategies for PDT cost-effectiveness, supporting value-based treatment and better patient outcomes.

Need for specialized treatments and therapies

Another factor that contributes to the high cost of PM is the need for specialized treatments and therapies. PM involves targeting specific genetic or molecular abnormalities that drive disease progression, which means that therapies must be developed and tested for each specific abnormality.

Once the underlying molecular/genetic etiology of a patient's condition has been determined, new insights into the underlying fundamental biology and disease pathogenesis may be gained using this knowledge. This will eventually help the development of medicines that specifically target the underlying cause of the disease. Development of the medicine will then be done for those patient populations most likely to benefit [10]. This will inevitably lead to a shift away from the development of "one-size-fits-all" medicines toward focused treatments that are more effective in limited patient groups. PM therapies will also necessitate the codevelopment of diagnostic tools to determine the optimal treatment for specific patients, driving the move from a stochastic clinical practice model. This requires significant investment in research and development, clinical trials, and regulatory approval. As a result, the cost of PM therapies can be much higher than traditional therapies, which are often developed for broad indications and based on more general mechanisms of action.

As much as US $350,000 per patient per year has been quoted as the cost of such targeted therapy. Despite their hefty price tags, some of these targeted medicines have shown, at least in small, restricted clinical studies, to drastically enhance survival, while others have only showed marginal improvement [11]. For example, in 2017, the U.S. Food and Drug Administration (FDA) approved a cancer drug called Kymriah (Tisagenlecleucel), which uses a patient's own genetically modified cells (T cells) to attack cancer cells (in B-cell acute lymphoblastic leukemia). The cost of Kymriah was initially set at $475,000 per treatment, making it one of the most expensive drugs ever approved by the FDA [12]. While the price has since been reduced, the high cost of PM drugs remains a significant financial challenge for patients and payers. While these drugs can be highly effective, they are often more expensive than traditional therapies.

Another example would be the case of spinal muscular atrophy, where there have been significant advancements in the provision of personalized care/therapeutics based on the specific mutations involved in the pathogenesis. However, these novel drugs themselves have high prices that are undoubtedly burdensome in many nations, especially low-middle income countries. The first medication authorized, nusinersen, costs around $750,000 in the first year of therapy and $375,000 per year after that. A single dosage of Zolgensma (Onasemnogene abeparvovec) has a lifetime cost of $2.1 million, and only a little over 2000 patients have received doses to date around the globe. Similarly, the weight-based cost of risdiplam might be as high as $340,000 per year [13]. This is in addition to the substantial costs involved with the routine treatment, resulting from the screen pro-gram implementations to the actual drug dosing, as well as costs associated with short- and long-term monitoring. Therefore, there is a stringent need to address the implementation gap in advanced therapeutics, from mutation-specific rare inherited disorders to common conditions such as cancers and seizures.

Costs associated with research and development, clinical trials, and regulatory approval

PM R&D is currently more expensive than standard treatment since it necessitates companion diagnostics and genetic testing. Companion diag-nostics frequently includes testing on biomarkers and marker-negative indi-viduals, necessitating bigger patient pools and higher costs. Furthermore, the requirement to establish evidence for PM entails the collecting of both real-world and nonclinical data. Unfortunately, present medical privacy and research policies have not matured to allow for the collecting of enor-mous volumes of data on big patient populations, and commercial health-care systems lack the technology to collect data on a large scale, raising the prices even higher. Patients suffer in the end because they carry a greater percentage of medical expenditures in the form of increasing out-of-pocket charges.

PM has therefore been less fully integrated into the traditional aspects of healthcare. According to some studies, only 40% of patients are aware of PM, and only 11% stated that their doctor mentioned PM as a treat-ment option with them [14]. However, the relative efficacy of PMs against standard methods has the potential to lower long-term expenditures.

Another factor contributing to the financial burden of PM is the challenge of ensuring patient access to personalized treatments. PM relies on identifying genetic variations that are associated with specific diseases or drug responses, but not all patients have equal access to genetic testing or personalized treatments [15]. This can lead to health disparities, with some patients receiving suboptimal care due to lack of access to PM. For example, a study published in the Journal of the American Medical Association in 2018 found that African American patients with breast can-cer were less likely than White patients to receive genetic testing, even after controlling for factors such as insurance coverage and clinical charac-teristics. This highlights the need for greater efforts to ensure equitable access to PM for all patients [16].

PM is still a relatively new field, and there is a lack of clear regulatory guidance on how to develop and approve PM treatments. This uncertainty can lead to increased costs for drug developers as they navigate the regulatory landscape. PM would benefit from agile governance processes to improve government coordination, leverage public–private partnerships, and actualize the potential of personalized healthcare without further exacerbating health inequity [17]. Two distinct approaches may be used to implement agile governance procedures in order to produce a more flexible and well-coordinated approach to healthcare regulation. A "design method" can recognize and handle issues as they appear, for instance, changing the design of a clinical study during a pandemic. A "system method" on the other hand considers the entire system and deals with structural issues. It can develop strong frameworks to address a variety of problems, including as building cross-national regulatory body systems to approve individualized medicines more swiftly and effectively, taking into account cost mitigation measures at every step.

The financial burden on patients

The high cost of PM can place a significant financial burden on patients, particularly those with rare or complex diseases that require specialized treatments. Patients may be required to pay out of pocket for genomic sequencing, biomarker testing, and other diagnostic procedures, which can add up to thousands of dollars. Even if these costs are covered by insurance, patients may still face significant out-of-pocket expenses for copays, deductibles, and other cost-sharing requirements.

In addition to the direct costs of PM, patients may also face indirect costs, such as lost wages, travel expenses, and other expenses associated with seeking specialized care. Patients with rare or complex diseases may need to travel long distances to access specialized care, which can be expensive and time-consuming [18]. They may also need to take time off work or arrange for childcare, which can further increase their financial burden. This can in turn impact the financial burden on patient outcomes and quality of life. This would be more concerning when the genomic sequences reveal surprised results that were not expected or those that were not really tested for. It can be mentally daunting to deal with additional diseases or conditions that patients can be at risk for, which would also demand screening and management of its own, adding on to the healthcare expenditure, for the patients and the caregivers as well.

There is no agreement on a price point for each additional year of quality-adjusted life (QALY). The willingness-to-pay threshold that is most frequently employed in the USA is 50,000 USD per QALY. This figure was traditionally based on the cost-effectiveness of dialysis from the 1970s, but it is now 130,000 USD per QALY. While the WHO suggests a willingness-to-pay threshold of three times the nation's GDP, the cost-effectiveness of PM remains unknown because of the numerous factors that determine it and the various willingness-to-pay criteria used [19]. As a result, we may need to take a new approach to valuing PM interventions.

The financial burden on payers and healthcare systems

The high cost of PM also places a significant financial burden on payers, such as private insurers, Medicare, and Medicaid, as well as on healthcare systems. Payers are responsible for covering the cost of PM therapies, which can be much higher than traditional therapies. This can lead to higher insurance premiums and out-of-pocket expenses for patients, as well as higher costs for employers and taxpayers who fund government-sponsored insurance programs [5].

Healthcare systems also bear the burden of providing specialized care and infrastructure to support PM. This includes investing in genomic sequencing and biomarker testing equipment, hiring specialized staff, and developing new treatments and therapies. These investments can be costly and may take years to yield a return on investment. In addition, healthcare systems may need to invest in new data management and analysis tools to handle the vast amounts of data generated by PM, which can also be expensive [15].

PM's integration into clinical practice has been emphasized as a priority for national and global health policy. Addressing the needs of the payer or reimbursement agency for healthcare is a major impediment to this development [20]. The economic argument for PM in this situation ultimately depends on proving its worth to those who make decisions about how to spend limited resources on healthcare. One way to provide data that might guide choices about how to allocate resources for healthcare is to compare pertinent options in terms of their costs and effects within an economic evaluation.

Instead of being a one-time stand-alone effort, the economic analysis of PM and the creation of new evidence should be ongoing processes. Thus, assessing the effects of adopting a diagnostic test in clinical practice on patients' long-term costs and health outcomes requires both decision-analytic modeling and prospective data gathering [21]. In the face of conflicting gaps in the evidence base for PM, value of information assessments may be used to rank research programs according to their potential to lower decision uncertainty.

Strategies for addressing the financial burden of precision medicine

While there are worries about the expense of PM, deliberate techniques to cost reduction can be used. Traditional therapy for genetically related illnesses is only successful in about 60% of individuals. Furthermore, individuals without a genetic diagnosis exhibit 38% adherence to their treatment plan after 2 years, compared to 86% for those who do. This ineffectiveness, along with shorter treatment programs, results in billions of dollars in long-term waste.

To address the financial burden of PM, several strategies have been proposed. One approach is to increase insurance coverage for genetic testing and personalized treatments. In 2019, the Centers for Medicare and Medicaid Services (CMS) proposed a national coverage determination for FDA-approved tests that use next-generation sequencing to guide cancer treatment decisions. This would make it easier for patients to access genetic testing and personalized treatments without facing high out-of-pocket costs [22]. However, such coverage determinations may not be feasible for all types of genetic testing and personalized treatments and may still require significant investments in healthcare infrastructure.

Another strategy to address the financial burden of PM is to improve the efficiency of drug development and approval. PM drugs often require extensive clinical trials and regulatory review, which can add to the cost and time required to bring new treatments to market. To address this, the FDA has implemented several initiatives aimed at streamlining the drug development and approval process for PM. For example, the FDA's Breakthrough Therapy Designation program is designed to expedite the development and review of drugs that show promise in treating serious or life-threatening conditions [23]. These programs have been used to accelerate the approval of several PM drugs.

Implications for the future of healthcare

cWGS can improve patient-care outcomes while lowering or remaining cost neutral. As the price of cWGS continues to fall, this will become increasingly in favor of cost reduction. In areas where specialists are scarce and clinical resources are underserved, it is still critical to provide equitable access to genomic medicine and, as a result, leapfrog these populations into next-generation care. Without addressing the issue of millions of undiagnosed individuals burdened by the diagnostic odyssey, genetic illness may become the leading cause of death in children in underresourced areas in less than a decade. Genetic health programs will need to significantly address genetic illnesses in LMICs, with exceptional outcomes for the undiagnosed. There will be a continued need for precision reimbursement for PM interventions: the need for patient-level decisions between payers, providers, and pharmaceutical companies that will also underline the future directions for research and policy in PM and healthcare finance.

References

[1] Gavan SP, Thompson AJ, Payne K. The economic case for precision medicine. Expert Rev Precis Med Drug Dev 2018;3(1):1−9. Available from: https://doi.org/10.1080/23808993.2018.1421858.

[2] Ree AH, Mælandsmo GM, Flatmark K, Russnes HG, Gómez Castañeda M, Aas E. Cost-effectiveness of molecularly matched off-label therapies for end-stage cancer − the MetAction precision medicine study. Acta Oncol 2022;61(8):955−62. Available from: https://doi.org/10.1080/0284186X.2022.2098053.

[3] Genetic Testing Registry [Internet]. [cited 2023 May 1]. https://www.ncbi.nlm.nih.gov/gtr/.

[4] Marshall DA, Grazziotin LR, Regier DA, Wordsworth S, Buchanan J, Phillips K, et al. Addressing challenges of economic evaluation in precision medicine using dynamic simulation modeling. Value Health [Internet] 2020;23(5):566−73. Available from: https://doi.org/10.1016/j.jval.2020.01.016.

[5] Geruso M, Jena AB, Layton TJ. Will personalized medicine mean higher costs for consumers? Harv Bus Rev [Internet] 2018;[cited 2023 May 1]. Available from: https://hbr.org/2018/03/will-personalized-medicine-mean-higher-costs-for-consumers.

[6] Lu CY, Terry V, Thomas DM. Precision medicine: affording the successes of science. NPJ Precis Oncol [Internet] 2023;7(1):3. Available from: https://doi.org/10.1038/s41698-022-00343-y.

[7] Phillips KA, Deverka PA, Hooker GW, Douglas MP. Genetic test availability and spending: where are we now? where are we going? Health Aff [Internet] 2018;37(5):710−16. Available from: https://doi.org/10.1377/hlthaff.2017.1427.

[8] Kris A, Wetterstrand MS. The cost of sequencing a human genome [Internet] Genomegov NHGRI 2019;[cited 2023 May 1]. Available from: https://www.genome.gov/about-genomics/fact-sheets/Sequencing-Human-Genome-cost.

[9] GoodSmith MS, Skandari MR, Huang ES, Naylor RN. The impact of biomarker screening and cascade genetic testing on the cost-effectiveness of MODY genetic testing. Diabetes Care [Internet] 2019;42(12):2247−55. Available from: https://doi.org/10.2337/dc19-0486.

[10] Dugger SA, Platt A, Goldstein DB. Drug development in the era of precision medicine. Nat Rev Drug Discov [Internet] 2018;17(3):183−96. Available from: https://doi.org/10.1038/nrd.2017.226.

[11] Shih Y-CT, Smieliauskas F, Geynisman DM, Kelly RJ, Smith TJ. Trends in the cost and use of targeted cancer therapies for the privately insured nonelderly: 2001 to 2011. J Clin Oncol [Internet] 2015;33(19):2190−6. Available from: https://doi.org/10.1200/JCO.2014.58.2320.

[12] Sagonowsky E. [No title] [Internet]. 2017 [cited 2023 May 1]. https://www.fierce-pharma.com/pharma/at-475-000-per-treatment-novartis-kymriah-a-bargain-or-just-another-example-skyrocketing.

[13] Leon-Astudillo C, Byrne BJ, Salloum RG. Addressing the implementation gap in advanced therapeutics for spinal muscular atrophy in the era of newborn screening programs. Front Neurol [Internet] 2022;13:1064194. Available from: https://doi.org/10.3389/fneur.2022.1064194.

[14] Johnson KB, Wei W-Q, Weeraratne D, Frisse ME, Misulis K, Rhee K, et al. Precision medicine, AI, and the future of personalized health care. Clin Transl Sci [Internet] 2021;14(1):86−93. Available from: https://doi.org/10.1111/cts.12884.

[15] Ferkol T, Quinton P. Precision medicine: at what price. Am J Respir Crit Care Med [Internet] 2015;192(6):658−9. Available from: https://doi.org/10.1164/rccm.201507-1428ED.

[16] Cutler DM. Early returns from the era of precision medicine. JAMA [Internet] 2020;323(2):109−10. Available from: https://doi.org/10.1001/jama.2019.20659.

[17] Doxzen KW, Signé L, Bowman DM. Advancing precision medicine through agile governance [Internet] Brookings. 2022[cited 2023 May 1]. Available from: https://www.brookings.edu/research/advancing-precision-medicine-through-agile-governance/.

[18] Koleva-Kolarova R, Buchanan J, Vellekoop H, Huygens S, Versteegh M, Mölken MR, et al. Financing and reimbursement models for personalised medicine: a systematic review to identify current models and future options. Appl Health Econ Health Policy [Internet] 2022;20(4):501−24. Available from: https://doi.org/10.1007/s40258-021-00714-9.

[19] Kasztura M, Richard A, Bempong N-E, Loncar D, Flahault A. Cost-effectiveness of precision medicine: a scoping review. Int J Public Health [Internet] 2019;64(9):1261−71. Available from: https://doi.org/10.1007/s00038-019-01298-x.

[20] Ginsburg GS, Phillips KA. Precision medicine: from science to value. Health Aff [Internet] 2018;37(5):694−701. Available from: https://doi.org/10.1377/hlthaff.2017.1624.

[21] Khoury MJ, Bowen S, Dotson WD, Drzymalla E, Green RF, Goldstein R, et al. Health equity in the implementation of genomics and precision medicine: a public health imperative. Genet Med [Internet] 2022;24(8):1630−9. Available from: https://doi.org/10.1016/j.gim.2022.04.009.

[22] CMS finalizes coverage of Next Generation Sequencing tests, ensuring enhanced access for cancer patients [Internet]. [cited 2023 May 1]. https://www.cms.gov/newsroom/press-releases/cms-finalizes-coverage-next-generation-sequencing-tests-ensuring-enhanced-access-cancer-patients.

[23] Office of the Commissioner. Breakthrough Therapy [Internet]. U.S. Food and Drug Administration. FDA; [cited 2023 May 1]. https://www.fda.gov/patients/fast-track-breakthrough-therapy-accelerated-approval-priority-review/breakthrough-therapy.

Index

Printed in the United States
by Baker & Taylor Publisher Services